SOTHIS PRESS

Sothis Press aims to completely reimagine the public domain book with an emphasis on universal access to captivatingly designed print editions. We believe that technology will enable every single book ever published to be perpetually accessible in beautiful print form. Sothis is beauty in ubiquity.

www.sothis.press
1 Waterhouse Square, London, ECIN 2ST

Arithmetic by Grades

William A. Campbell

SOTHIS PRESS

ISBN 978-1-0365-1936-0

This edition is published by
Sothis Press © 2023. The company
is registered in England and Wales.
The Company Number is 14900389.

Hinds, Hayden & Eldredge Textbooks by Grades

ARITHMETIC BY GRADES

BOOK ONE

BY

WILLIAM A. CAMPBELL
DISTRICT SUPERINTENDENT OF SCHOOLS
NEW YORK CITY

AND

THOMAS H. HUGHES
EVANDER CHILDS HIGH SCHOOL
NEW YORK CITY

HINDS, HAYDEN & ELDREDGE, Inc.
New York Philadelphia Chicago

PREFACE

THIS book, the first in a new series of practical arithmetics, is designed to impart to the pupils a mastery of the fundamental processes in manipulating integers, familiarity with a few rudimentary fractions, and a facility in simple calculation. It is hoped that the method, and particularly the manner, of presentation will not only guide but also interest the pupils in the ordinary applications of whole numbers and simple fractions to everyday problems.

The authors have never lost sight of the fact that a thorough grounding in the fundamental operations is the key to future success in arithmetical computation, and that proficiency in the use of tables and factors must be gained through drill. Accordingly the work in problems in the earlier books of the series has been subordinated to drill practice. It is hoped that sufficient and varied drill work has been provided so that abstract calculation may become almost automatic, and the pupil may acquire that accuracy and rapidity which are necessary if the thought work of the later years is to be unhampered. The fact, however, that all work in the abstract should follow as a result of a concrete and urgent need in the pupil's mind, that there must be a vital point of contact between the pupil and the new topic or process, has been kept in view and, as a rule, all drill work and new processes are approached in problem form.

An effort has been made also to adapt the material to the interests and experiences of the pupils and to keep the wording of the text well within their comprehension. The authors have attempted to embody in the series the results growing out of the experience and the criticism of the most successful teachers of the past few years, and to follow the spirit of the courses of study of the most progressive cities and states, especially in the matter of the elimination of obsolete methods and topics. They have tried to reduce the explanations of formal processes to the most economical and complete forms for use in the study of new lessons. In the wording of problems and in the avoidance of any routine order of presentation or questioning, they have tried constantly to throw the pupil upon his own resources. It is of vital importance that pupils be taught as early as possible the intelligent use of a textbook, for the book is to become the chief source of knowledge when school days are over. An effective text should be planned so as not merely to supply series of exercises or problems, but also to enable a teacher to instruct pupils in the art of acquiring knowledge from the book, and in this way to foster mental independence, develop self-reliance, and promote self-instruction.

CONTENTS

THIRD YEAR BOOK

FIRST HALF: GRADE 3 A

SECOND HALF: GRADE 3 B

ARITHMETIC BY GRADES

THIRD YEAR BOOK

FIRST HALF: GRADE 3A

I. READING AND WRITING NUMBERS

ORAL EXERCISE

Read these numbers:

1.	2.	3.	4.	5.
14	175	301	416	249
29	222	416	728	330
84	51	118	837	561
308	615	220	342	161

thousands	hundreds	tens	units
1	3	4	2

Read: One thousand three hundred forty-two. (Do not use the word *and.*) In 1342 there are 1 thousand, 3 hundreds, 4 tens, and 2 units.

Read these numbers:

6.	7.	8.	9.	10.
3,500	7,482	2,184	1,324	4,842
9,620	1,775	5,376	3,985	1,183
3,370	7,769	4,986	2,474	9,607

WRITTEN EXERCISE

Write these numbers in figures:

1. Three hundred forty-two.
2. Five hundred fifty-five.
3. Six hundred thirteen.
4. Eight hundred ninety-five.
5. Four hundred ten.
6. Seven hundred fourteen.
7. Nine hundred sixty-seven.
8. Nine hundred ninety-one.

Write 1342, *and the following numbers under it:*

9. Five thousand.
10. Six thousand six hundred.
11. One thousand five hundred.
12. Seven thousand one hundred.
13. Three thousand.
14. Two thousand two hundred.
15. Eight thousand seven hundred.
16. Nine thousand three hundred.

Dollars and Cents.

Read:	**$1**	**$7**	**$9**	**$4**	**$2**
	$11	**$27**	**$19**	**$44**	**$25**

We write:	Five cents	$.05	Seven cents	$.07
	Ten cents	$.10	Eleven cents	$.11
	Fifteen cents	$.15	Nineteen cents	$.19
	Twenty-five cents	$.25		

We may write also: 5¢, 10¢, 15¢, 7¢, 11¢, 19¢, 25¢.

We write: One dollar and five cents $1.05
Two dollars and twelve cents $2.12

Notice the decimal point separating dollars and cents. In reading dollars and cents, we read the decimal point as **and**.

ORAL EXERCISE

Read:

1.	2.	3.	4.
$.12	$12.00	$.02	$38.35
$.18	$15.10	$12.17	$20.98
$4.07	$19.22	$.88	$5.75
$162.75	$435.40	$892.67	$990.56

WRITTEN EXERCISE

Write these amounts:

1. Five dollars.
2. Twenty-five dollars.
3. Two dollars and ten cents.
4. Seventeen dollars and two cents.
5. Eight dollars and eighty cents.
6. One thousand dollars.
7. Thirteen dollars and eighty-eight cents.
8. Four hundred fourteen dollars and three cents.
9. Three thousand five hundred one dollars.
10. Nine hundred nineteen dollars and fifty-six cents.
11. Five thousand four hundred eighty dollars.
12. Two hundred eighteen dollars and twelve cents.
13. Six thousand eight hundred dollars and ten cents.
14. Three hundred dollars and six cents.

Roman Numbers.

1. Find these figures on a clock:

one	two	three	four	five	six
I	II	III	IV	V	VI
seven	eight	nine	ten	eleven	twelve
VII	VIII	IX	X	XI	XII

2. Learn:

XIII	XIV	XV	XVI
13	14	15	16
XVII	XVIII	XIX	XX
17	18	19	20

ORAL EXERCISE

Read:

1. VI	VIII	X	XIV	XV	XVIII
2. IV	XII	IX	XI	XVII	XVI
3. V	XX	III	XIII	VII	IX

WRITTEN EXERCISE

Write the Roman numbers for:

1. six	2. four	3. nine
fifteen	twelve	thirteen
eleven	twenty	five
4. three	5. eight	6. eighteen
seventeen	fourteen	seven
ten	nineteen	sixteen

II. COUNTING

I. Count:

1. By 2's from 3 to 39.
2. By 3's from 7 to 37.
3. By 4's from 9 to 53.
4. By 5's from 2 to 57.
5. By 4's from 7 to 51.
6. By 3's from 2 to 35.
7. By 2's from 5 to 33.
8. By 5's from 4 to 64.

II. 1. Beginning with 44, count by 2's to 8.
2. Beginning with 41, count by 3's to 5.
3. Beginning with 53, count by 4's to 5.
4. Beginning with 66, count by 5's to 6.

III. 1. From 51, count by 3's to 6.
2. From 55, count by 4's to 7.
3. From 64, count by 5's to 4.
4. From 45, count by 2's to 3.

IV. Count:

1. By 3's from 2 to 41.
2. By 4's from 3 to 51.
3. By 2's from 5 to 45.
4. By 5's from 6 to 76.
5. By 3's from 7 to 52.
6. By 4's from 8 to 68.
7. By 5's from 9 to 79.
8. By 5's from 7 to 97.
9. By 2's from 8 to 38.
10. By 3's from 9 to 39.

V. Count:

1. By 6's from 6 to 66.
2. By 6's from 9 to 75.
3. By 6's from 5 to 77.
4. By 6's from 7 to 73.
5. By 6's from 6 to 78.
6. By 6's from 4 to 82.

III. ADDITION

ORAL EXERCISE

1. Henry has 6 cents and John has 7 cents. How much have they together?

2. I have 11 picture cards and you have 4. How many have we together?

The answer in addition is called the **sum.**

3. Kate spent 7 cents; her sister spent 4 cents. Find the sum they spent together.

4. Find the sum of 9 and 7.

5. What sum shall I have if I add 8 and 4?

6. What sum shall I have if I add 14 to 6?

7. Find the sum of 12 and 5.

Give the sums rapidly:

8.	6	5	7	8	9	2	3	5	7	8	9	6
	4	6	8	7	3	9	8	7	6	5	4	9

9.	14	29	76	47	59	67	88	36	79
	2	2	2	2	2	2	2	2	2

10.	67	58	27	83	48	33	77	68	97
	3	3	3	3	3	3	3	3	3

11. Add in turn the numbers from 1 to 10 to the numbers in Ex. **8.**

12. Add 4 and 5 in turn to the numbers in **9** and **10.**

Terms in Addition. The numbers to be added in addition are called the **addends.** The result is called the **sum.**

We check the work in addition by adding in the opposite direction. If we first add the columns down, we then add them up.

WRITTEN EXERCISE

Add and check:

1.	**2.**	**3.**	**4.**	**5.**	**6.**	**7.**
44	26	37	40	59	66	72
60	94	83	72	60	54	65
73	40	54	67	73	80	90
52	42	63	57	71	80	72

8.	**9.**	**10.**	**11.**	**12.**	**13.**	**14.**
28	60	36	54	89	45	72
90	54	80	65	36	70	40
87	22	65	40	54	11	76
47	53	73	61	20	72	84

15.	**16.**	**17.**	**18.**	**19.**	**20.**	**21.**
34	42	27	36	45	54	63
46	97	72	81	90	89	78
50	20	66	54	63	72	81
27	83	20	30	41	50	62

22.	**23.**	**24.**	**25.**	**26.**	**27.**	**28.**
63	46	97	88	76	65	54
47	57	43	32	21	40	30
59	38	78	87	66	65	74
21	22	80	30	41	22	38

ORAL EXERCISE

Give the sums rapidly:

1.	9	8	7	6	7	8	9	9	8	7	6	7
	9	7	8	7	6	9	8	7	8	9	7	6

2.	57	96	56	97	68	77	89	36	76
	5	5	5	5	5	5	5	5	5
3.	14	77	34	97	57	77	25	86	68
	6	6	6	6	6	6	6	6	6

WRITTEN EXERCISE

Add and check:

1.	**2.**	**3.**	**4.**	**5.**	**6.**	**7.**
57	64	43	41	24	78	94
21	11	32	43	55	66	78
96	87	73	54	67	76	83
84	72	63	54	41	30	12

8.	**9.**	**10.**	**11.**	**12.**	**13.**
346	722	463	514	146	731
539	178	229	357	448	239

14.	**15.**	**16.**	**17.**	**18.**	**19.**
832	547	611	442	286	863
139	448	379	548	707	127

20.	**21.**	**22.**	**23.**	**24.**	**25.**	**26.**
57	54	63	92	84	76	52
87	63	72	84	76	63	79
52	75	69	23	34	46	50
12	63	13	42	17	21	19

ORAL EXERCISE

Add:

1.	33	79	25	56	84	47	68
	9	9	9	9	9	9	9

2.	15	48	59	26	61	72	37
	8	8	8	8	8	8	8

3. Add 7 and 8 in turn to the numbers in Ex. **1.**

4. Add 7 and 9 in turn to the numbers in **2.**

Give these sums rapidly:

5.	2	3	5	4	3	7	8	6	3
	4	3	2	9	8	9	7	6	7
	8	5	7	4	4	2	3	5	9

6.	2	3	5	2	4	3	2	1	5
	7	3	2	8	5	6	7	8	7
	3	5	7	3	3	4	4	5	4

Add:

7.	9	7	8	6	5	7	9	6	4
	4	2	3	4	4	3	2	4	6
	8	8	6	5	4	9	7	6	9

8.	5	4	3	4	4	5	5	8	5
	9	8	7	6	6	7	9	3	7
	4	5	3	8	9	9	7	9	1
	5	3	2	4	1	3	2	4	3

WRITTEN EXERCISE

1. Add 23, 158, 42, and 558:

23 158 42 558 781	The sum of 8, 2, 8, and 3 is 21. We write 1 under units' column and carry 2 to the tens' column. The sum of 7, 4, 5, and 2 is 18. We write 8 in the tens' column and carry 1 to the hundreds' column. The sum of 6 and 1 is 7. We prove that the sum is correct by adding each column again, in the opposite direction; if we first add the columns up, we then add them down.

Add the following and test the sums:

2. 54	**3.** 17	**4.** 228	**5.** 8	**6.** 547
76	54	337	24	66
105	210	16	336	122
8	562	22	417	88

7. 29	**8.** 119	**9.** 221	**10.** 42	**11.** 45
38	27	69	633	501
417	76	65	102	24
26	259	347	262	18
5	92	11	14	71

12. 18	**13.** 552	**14.** 841	**15.** 675	**16.** 781
254	347	561	28	482
963	63	384	924	174
417	148	78	602	86
544	901	397	88	392

Adding Dollars and Cents. Note that in adding dollars and cents, we write the decimal points underneath one another. We write the dollar sign before the first number and before the sum.

Add and test:

1.	2.	3.	4.	5.
$ 4.76	$ .68	$ .45	$ 3.25	$19.22
44.22	7.19	.82	13.17	16.04
3.08	.29	22.00	6.75	7.25
5.15	14.03	17.79	3.45	17.11
2.25	.75	4.25	4.80	9.22

6.	7.	8.	9.	10.
$24.56	$30.18	$20.04	$ 5.46	$ .09
1.75	20.00	.61	72.24	3.04
9.84	5.96	.79	.28	.16
13.88	7.84	82.28	3.49	15.05
5.54	18.88	6.14	.16	.77

Write in figures and add:

11. Eleven dollars and twenty-two cents.
12. Five hundred seven dollars and twenty cents.
13. Six dollars and five cents.
14. Three hundred dollars and forty-four cents.
15. Fifty-nine dollars and nineteen cents.

WRITTEN EXERCISE

Add and check:

1.	2.	3.	4.
342	462	467	$ 2.78
587	1002	1008	28.07
432	984	672	6.74
682	1148	988	11.22

5.	502	**6.**	1586	**7.**	$ 45.61	**8.**	347
	11		374		66.72		1172
	362		1598		989.84		564
	1374		787		2.01		387
9.	1202	**10.**	989	**11.**	1872	**12.**	36
	488		1472		288		489
	984		674		54		1670
	2472		88		709		342
	986		864		1431		66
13.	$244.18	**14.**	414	**15.**	1166	**16.**	487
	728.47		738		344		52
	516.16		1292		718		975
	72.08		891		1001		343
	494.75		434		444		67
17.	52	**18.**	$ 72.14	**19.**	255	**20.**	$240.04
	174		162.10		602		6.00
	744		5.05		588		.76
	1302		705.40		1002		562.14
	500		22.22		375		783.84
21.	532	**22.**	$115.15	**23.**	$ 7.84	**24.**	1034
	614		6.01		3.22		2014
	1002		24.29		.09		82
	586		5.36		117.11		76
	1298		141.05		22.24		1001

Add and check:

25. 27,380	**26.** 16,712	**27.** 29,422	**28.** 24,609	**29.** 36,407
41,985	7,583	790	34,842	4,892
6,641	83,921	77,028	4,560	1,004
90,810	52,000	10,152	4,108	18,409
40,058	7,820	9,380	35,087	7,623
2,000	97,645	10,504	17,080	15,603
12,597	9,380	1,089	5,897	5,891

30. 65,977	**31.** 4,108	**32.** 14,309	**33.** 5,680	**34.** 84,587
4,585	34,560	33,642	93,040	3,852
47,725	5,687	4,380	4,872	60,971
7,600	22,409	11,259	64,321	30,484
34,382	92,928	64,328	2,575	16,873
84,484	62,937	17,289	84,300	8,298
5,683	37,735	10,025	12,470	44,105

WRITTEN REVIEW PROBLEMS

1. In the first car of a train there are 58 passengers, in the second car 44, in the third car 67, in the fourth car 59. How many passengers on the train?

2. There are 45 pupils present in one class, 48 in a second, 51 in a third, and 39 in a fourth. How many pupils are present in the four classes?

3. A clerk took in $144.06 on Monday, $58.92 on Tuesday, $125 on Wednesday, and $89.43 on Thursday. How much did he receive in all?

4. A telephone rang 117 times in one week, 143 times in another, 98 times in a third, and 127 times in a fourth week. How many calls were rung?

5. There are 142 pages in a reader, 92 pages in a speller, 126 pages in a history, and 138 in an arithmetic. How many pages in all these books?

6. The monitor put 92 books on one shelf, 56 on another shelf, 118 on a third shelf, and 102 on a fourth shelf. How many books were placed on all the shelves?

7. The milkman delivered 185 quarts milk the first week, 203 the second week, 196 the third week, and 218 the fourth week. How many quarts did he deliver in all?

8. A grocer sold 3 lb. coffee, $.96; 2 lb. butter, $.64; 1 mop, $1.20; and 3 loaves bread, $.15. Find the cost of all these articles.

9. A dealer sold a couch for $25.40, 3 chairs for $7.30, and a picture for $8.65. How much did all the articles cost?

10. There are 389 pupils on the first floor, 593 on the second, 702 on the third, and 1116 on the fourth. How many pupils are there in the building?

11. A storekeeper sold goods on Monday amounting to $428.16, on Tuesday $503.37, on Wednesday $408.05, on Thursday $377.75. How much did his sales amount to for the four days?

IV. SUBTRACTION

ORAL EXERCISE

1. Sarah had 9 cents and she spent 4 cents. How much had she left? What number added to 4 will make nine?

2. I have saved 3 cents to buy a notebook. The book costs 7 cents. How much more must I save? What number added to 3 will make 7?

3. There are 8 books on my desk. 5 are in one pile. How many are in the other? What number added to 5 will make 8?

4. Harry is 8 years old. In how many years will he be 13 years old? What number added to 8 will make 13?

5. I bought a pencil box for 12 cents. I gave the man a dime and a nickel. How much change shall I get?

6. The grocer had 14 bananas in a bunch. He sold 6. How many has he left?

7. George walked 12 blocks to the museum. How many more blocks did he walk than Harry, who walked only 8?

8. Mother bought 3 oranges for 9 cents. She gave the man a quarter. How much change did she receive?

Subtracting Numbers of Two Figures. There were 32 books in the closet. A boy brought out 18. How many are left?

32 18 — 14	There is no number which added to 8 will make 2. We add 10 to 2, making 12. When 10 is added to a figure in the minuend, 1 must be added to the next figure in the subtrahend. What number added to 8 will make 12? 8 and 4 are 12. Write 4, add 1 to 1, making 2. What number added to 2 will make 3? 2 and 1 are 3. Write 1.

WRITTEN EXERCISE

Subtract:

1. 22	**2.** 34	**3.** 53	**4.** 72	**5.** 91	**6.** 83	**7.** 41
10	16	37	48	63	47	17
8. 51	**9.** 43	**10.** 54	**11.** 72	**12.** 90	**13.** 63	**14.** 70
37	28	18	49	54	38	29

Terms in Subtraction.

68 *minuend* 39 *subtrahend* — 29 *difference* or *remainder* Proof: 39+29=68	The larger number in subtraction is called the **minuend.** The smaller number is called the **subtrahend.** The answer is called the **difference** or **remainder.**

We check the work in subtraction by adding the difference and the subtrahend.

WRITTEN EXERCISE

Subtract and check:

1. 63	**2.** 84	**3.** 33	**4.** 71	**5.** 95	**6.** 60	**7.** 24
48	37	19	29	36	44	15

8. 67	**9.** 74	**10.** 82	**11.** 83	**12.** 96	**13.** 47	**14.** 67
39	28	43	17	18	29	38

ORAL EXERCISE

Subtract:

1. 52	63	92	22	41	74	35	85
4	5	8	9	5	6	8	7

2. Subtract in turn all the numbers from 2 to 10 from the numbers in Ex. **1.**

Subtract:

3. 23	52	74	41	91	82	32	63
4	4	4	4	4	4	4	4

4. Subtract 5 and 6 in turn from the numbers in **3.**

Supply the subtrahends:

5. 50¢	18¢	24¢	35¢	46¢	17¢	33¢	42¢
?	?	?	?	?	?	?	?
40¢	11¢	13¢	20¢	22¢	9¢	17¢	22¢

6. 32	76	65	43	27	94	72	85
?	?	?	?	?	?	?	?
10	52	41	26	11	75	26	34

Supply the minuends:

7.	?	?	?	?	?	?	?	?
	12	8	4	17	8	18	3	8
	30	17	11	22	18	35	14	27
8.	?	?	?	?	?	?	?	?
	21¢	24¢	51¢	33¢	24¢	38¢	12¢	43¢
	54¢	67¢	60¢	42¢	76¢	37¢	83¢	29¢

ORAL EXERCISE

Subtract:

1.	76	54	35	27	85	42	91	63
	8	8	8	8	8	8	8	8
2.	47	22	63	84	35	54	92	76
	9	9	9	9	9	9	9	9

3. Subtract 6, 7, and 9 from the numbers in Ex. 1.
4. Subtract 6, 7, and 8 from the numbers in Ex. 2.

ORAL PROBLEMS

1. May had 8 picture cards. She found 13 more. How many has she now?

2. Last month she had 32 cards and gave away 9. How many had she then?

3. Her father gave her 6, her uncle 12, and her sister 5. How many did she then have?

4. Ethel bought a bottle of milk for 9¢ and 2 loaves of bread at 5¢ each. What did she spend?

5. How much change did she get from a quarter?

6. Mother bought 36 eggs and then found 8 in the closet. How many had she in all?

7. John had 18 spelling words correct. Grace had 7 more correct. How many had Grace correct?

8. How much change shall I get from a dollar bill if I pay 85¢ for medicine?

WRITTEN PROBLEMS

1. We had 82 notebooks in one closet. We sent 49 to Miss Mill. How many have we now?

2. The newsboy had 84 papers and he sold 39. How many has he left?

3. My mother gave 49¢ to the butcher, 12¢ to the baker, and 17¢ to the grocer. How much did she spend?

4. The grocer had 56 cans of corn. He sold 37. How many has he now?

5. George had 37 marbles and he won 19 more. How many had he then?

6. Sam paid 8¢ for a pencil box, 4¢ for a strap, and 9¢ for a blank book. How much did he have left from a quarter?

7. Helen bought 19 two-cent stamps for her father. How much change did she get from a half dollar?

8. Our class used 18 pieces of chalk. Now there are 59 pieces left in the box. How many were there in the box at first?

9. A carpenter had a plank 41 feet long. He cut out 18 feet. How long is the plank now?

10. Gertrude and John together weigh 99 pounds. John weighs 51 pounds. What is Gertrude's weight?

11. We gave out 43 readers. Now there are 28 left in the closet. How many readers have we in all?

ORAL EXERCISE

Subtract:

1.	58	76	95	36	84	26	67	45
	7	7	7	7	7	7	7	7
2.	18	68	33	58	86	77	48	29
	9	9	9	9	9	9	9	9
3.	93	77	68	85	57	86	45	76
	8	8	8	8	8	8	8	8

4. Subtract 4 and 5 in turn from each number above.

Find the difference:

5.	22	65	71	43	82	34	91	53
	4	4	4	4	4	4	4	4
6.	91	54	41	20	82	32	60	71
	7	7	7	7	7	7	7	7

7. Subtract 9 and 8 in turn from each number above.

Subtracting Numbers of Three Figures.

Subtract 152 from 375:

	What number added to 2 will make 5?
375	2 and **3** are 5. Write 3.
152	5 and **2** are 7. Write 2.
	1 and **2** are 3. Write 2.
223	Notice that we write units under units, tens under tens, and hundreds under hundreds.

Proof: Add 223 and 152 without rewriting. If your work is correct, the sum should equal the minuend.

WRITTEN EXERCISE

Subtract and check:

1. 785 472	**2.** 957 685	**3.** 488 434	**4.** 769 526	**5.** 596 382	**6.** 949 553
7. 637 525	**8.** 452 141	**9.** 985 772	**10.** 428 312	**11.** 268 140	**12.** 149 124
13. 128 107	**14.** 643 131	**15.** 293 172	**16.** 142 120	**17.** 583 481	**18.** 672 461
19. 836 425	**20.** 547 206	**21.** 678 364	**22.** 845 703	**23.** 326 126	**24.** 739 527
25. 868 343	**26.** 735 514	**27.** 289 172	**28.** 247 135	**29.** 583 471	**30.** 946 346

WRITTEN EXERCISE

1. Subtract 382 from 541:

	There is no number which added to 2 will make 1.
	We add 10 to 1, making 11.
541	When 10 is added to a figure in the minuend,
382	1 must be added to the next figure in the sub-
159	trahend.
	2 and **9** are 11. Write 9. Add 1 to 8.
	9 and **5** are 14. Write 5. Add 1 to 3.
	4 and **1** are 5. Write 1.
Proof:	382+159=541.

Find and prove these remainders:

2. 542	**3.** 783	**4.** 634	**5.** 673	**6.** 542	**7.** 462
297	469	345	499	347	374
8. 243	**9.** 701	**10.** 542	**11.** 673	**12.** 211	**13.** 619
47	688	485	304	98	226
14. 964	**15.** 228	**16.** 416	**17.** 100	**18.** 285	**19.** 161
526	178	375	81	179	82
20. 240	**21.** 387	**22.** 460	**23.** 990	**24.** 986	**25.** 860
219	248	375	496	749	246
26. 740	**27.** 653	**28.** 890	**29.** 942	**30.** 760	**31.** 464
288	376	479	855	326	407

Subtracting Dollars and Cents. In subtracting dollars and cents, we write the decimal points under one another. We write the dollar sign before the minuend and the difference.

WRITTEN EXERCISE

Subtract:

1. $8.65	2. $4.62	3. $7.25	4. $9.00
4.53	3.74	5.09	7.15
$	$	$	$
5. $.76	6. $3.00	7. $.52	8. $52.61
.59	.85	.27	9.75
$	$	$	$
9. $82.00	10. $74.60	11. $20.02	12. $2.78
2.88	6.84	3.70	.99
$	$	$	$
13. $566.62	14. $281.90	15. $18.01	16. $35.40
28.84	153.28	5.68	19.70
$	$	$	$
17. $92.62	18. $12.79	19. $56.72	20. $84.40
8.88	3.89	42.72	14.01
$	$	$	$

21. From forty-two dollars and ten cents, subtract twenty-seven dollars and twelve cents.

22. From five thousand six hundred forty-two, subtract two thousand nine hundred fifty-six.

WRITTEN EXERCISE

Subtract:

1. $324.70	2. $554.64	3. $224.26	4. $330.40
218.82	49.85	119.47	111.52
5. $664.00	6. $532.00	7. $786.66	8. $865.42
230.08	26.24	239.87	343.89
9. $786.95	10. $523.44	11. $535.14	12. $845.25
228.98	417.28	227.20	346.35

13. $352.46 – $18.79 = 14. $667.88 – $52.39 =
15. $311.01 – $.16 = 16. $19.64 – $4.29 =
17. $226.24 – $85.19 = 18. $135.03 – $.98 =

Subtracting Numbers of Four Figures.
Subtract 3284 from 7900.

7900	4 and **6** make 10.	Write 6.	Add 1 to 8.
3284	9 and **1** make 10.	Write 1.	Add 1 to 2.
	3 and **6** make 9.	Write 6.	
4616	3 and **4** make 7.	Write 4.	
Proof:	3284+4616=7900.		

WRITTEN EXERCISE

Subtract and prove the difference:

1. 1893	2. 4441	3. 5430	4. $264.10
647	332	4672	153.92
5. $330.16	6. $410.14	7. $116.22	8. $424.41
211.18	39.95	15.83	243.53

Subtract and check:

9. 6567 3487	**10.** $811.39 272.59	**11.** $424.40 185.60	**12.** $254.10 135.52
13. 4750 17	**14.** 647 8	**15.** 5338 49	**16.** 6371 988
17. 5092 4108	**18.** 2142 1258	**19.** 8353 7983	**20.** 8470 7988
21. 5064 3689	**22.** 2989 1389	**23.** 4907 748	**24.** $833.02 722.98
25. 4000 2706	**26.** 6140 3462	**27.** 5000 3403	**28.** 7180 4293
29. 6000 86	**30.** 9240 8197	**31.** 4250 3282	**32.** 6350 4763

33. Subtract seven hundred thirty-three from one thousand one hundred sixteen.

34. From two thousand three hundred six, subtract one hundred eighty-nine.

35. From six thousand one hundred dollars and thirteen cents, take four thousand five hundred forty-eight dollars and eighty-seven cents.

36. Subtract nine hundred ninety-five dollars and twenty-nine cents from three thousand dollars.

ORAL DRILL EXERCISE

Add:

	A		B		C		D		E	
1.	46	23	54	46	32	87	82	18	94	36
	7	8	6	9	8	5	8	9	7	6
2.	5	6	9	7	7	8	8	6	9	5
	23	62	54	34	41	91	47	56	72	68
3.	2	5	4	9	7	1	3	8	5	6
	4	3	5	4	4	6	6	5	6	5
	7	8	9	6	8	9	7	6	9	8

Subtract:

	A		B		C		D		E	
4.	53	27	46	93	38	84	49	71	58	63
	−8	−9	−7	−6	−9	−5	−8	−6	−8	−9
5.	33	57	44	38	16	63	77	81	84	92
	−7	−8	−6	−7	−9	−5	−8	−4	−6	−3

Give answers:

	A	B	C	D	E
6.	2×3	$\frac{1}{2}$ of 10	2×8	$32-28$	$18+9$
7.	$4\div2$	$45+9$	$55-8$	8×5	$34-11$
8.	$\frac{1}{2}$ of 8	5×4	$?\div3=3$	$12\div3$	$15\div3$
9.	$34+7$	$17-9$	7×3	$54+12$	6×3
10.	$21-8$	$18\div2$	$25+11$	$12\div?=2$	$\frac{1}{2}$ of 12
11.	4×4	6×3	7×5	5×10	2×6
12.	$45+7$	4×5	$27-8$	6×4	$12\div2$

GENERAL ORAL PROBLEMS

1. Sadie's father gave her 8 cents for a pencil box and 5 cents for a strap. How much did she receive?

2. I bought 5 quarts of milk at 8 cents a quart. How much change must I take home out of 50 cents?

3. Harry bought 6 peaches at 3 cents each. How much change had he left from 20 cents?

4. John made 12 cents profit every day selling newspapers. How much will he make in six days?

5. Van had 45 cents. He spent 9¢. How many cents did he then have?

6. What change should I receive from a ten-dollar bill after paying $9.50 for a chair?

7. Gertrude had 18 words correct in the test of 25 words. How many had she wrong?

8. Ethel's father sold her bicycle for $18. He paid $25 for it last year. How much did he lose?

9. Van bought 4 two-cent stamps, 8 one-cent stamps and 4 postal cards. How much did he give the clerk?

10. Margaret bought 9 oranges at 3 cents each and gave the clerk 30 cents. How much change did she receive?

11. Frank had 28 words on his paper and wrote 7 more. How many words did he then have?

12. Louis bought milk for 3¢, potatoes for 6¢, and bread and butter for 5¢. What did his lunch cost him?

GENERAL WRITTEN PROBLEMS

1. I spent $8.50 in one store and $13.25 in another. How much did I spend in both stores?

2. Sam sold his bicycle for $18.75. It cost him $23.50. How much did he lose?

3. Harry's mother had $10 in her purse. She spent $.72 in the drug store, $1.84 in the bakery, and $4.65 in the butcher's. How much did she bring back?

4. There are 37 pupils in my class, 49 in Ethel's, 50 in Clara's, and 34 in Tom's. How many are there in the four rooms?

5. We had 45 paint brushes. We wore out 9, loaned 8, and lost 3. How many have we now?

6. Paul's uncle bought an automobile for $760. He sold it for $825. How much did he make?

7. Ely had $4.20. He spent $3.87. How much did he then have?

8. George had 49 letters to mail. He placed a two-cent stamp on each. How much did the postage cost him?

9. My mother spent $31.50 for clothing for my brother and me. My suit cost $15. What did his suit cost?

10. Kate's sister paid $4.50 for a hat, $18.00 for a dress, and $4.25 for shoes. How much did she spend?

11. Mrs. Harris spent $2.49 in the dry goods store and twice as much in the butcher's. How much did she spend in both stores?

V. MULTIPLICATION

Multiplying by 2.

1. How much will two sandwiches cost at 3 cents each? How much is two times three?

2. There are 7 dishes in each row. How many in two rows? How much is 2 times 7?

3. How much is 2 times 8? 2 times 6? 2 times 4? 2 times 9? 2×3? 2×2? 2×6?

4. Give these sums:

2	3	4	5	6	7	8	9	10	11	12
2	3	4	5	6	7	8	9	10	11	12

The Table of 2's. Learn this:

2×1=2	2×7 =14	1×2=2	7×2=14
2×2=4	2×8 =16	2×2=4	8×2=16
2×3=6	2×9 =18	3×2=6	9×2=18
2×4=8	2×10=20	4×2=8	10×2=20
2×5=10	2×11=22	5×2=10	11×2=22
2×6=12	2×12=24	6×2=12	12×2=24

1.

3	11	4	9	6	7	8	10	12
×2	×2	×2	×2	×2	×2	×2	×2	×2

2. How many are 4 times 2? 6 times 2? 3 times 2? 7 times 2? 2 times 8?

ORAL EXERCISE

Give answers rapidly:

1.	2.	3.	4.
5×2	1×2	11×2	2×10
7×2	2×7	2×11	2×8
8×2	9×2	2×12	2×9
4×2	2×3	6×2	2×3
3×2	4×2	2×5	2×1
2×7	3×2	2×4	2×6

5. Two 4's=? Two 9's=? Three 2's=? Ten 2's=?
6. Five 2's=? Six 2's=? Two 12's=? Two 9's=?
7. Seven 2's=? Two 10's=? Eight 2's=?

ORAL PROBLEMS

1. If pencils cost 2 cents each, how much will 8 pencils cost?
2. Find the cost of 4 apples at 2 cents each.
3. What will the two-cent stamps cost for 7 letters each requiring one stamp?
4. I lost two nickels. How many cents did I lose?
5. If a newspaper costs 2 cents a copy, what will 5 copies cost?
6. Harry has 2 rulers. Mary has twice as many. How many has Mary?
7. The monitor has 8 times as many as Harry. How many has he?
8. Find the cost of 2 baseballs at 6 cents each.

Multipliers of One Figure.

1. Add:	21	8	74	11	84	35	65	93	42
	21	8	74	11	84	35	65	93	42

In these examples we added each number to itself or **multiplied** it by 2.

2. Learn this quicker way of getting the answer.

Multiply 42 by 2:

42	2 times 2 units are 4 units. Write 4 under 2.
×2	2 times 4 tens are 8 tens. Write 8 under 4.
84	*Write units under units, and tens under tens.*

Copy these examples and find the products:

3. 41	4. 22	5. 22	6. 32	7. 24
2	3	4	3	2
8. 34	9. 31	10. 33	11. 43	12. 30
2	3	2	2	3

Terms in Multiplication. The sign of multiplication is ×. This sign is read "times" or "multiplied by."

The number to be multiplied, 42 in the example above, is called the **multiplicand**.

The number by which we multiply, 2 in the example above, is called the **multiplier**.

The result in multiplication is called the **product**.

WRITTEN EXERCISE

Multiply:

1. 21	**2.** 33	**3.** 42	**4.** 22	**5.** 54
2	2	2	2	2
6. 62	**7.** 23	**8.** 34	**9.** 44	**10.** 52
2	2	2	2	2
11. 30	**12.** 51	**13.** 41	**14.** 24	**15.** 112
2	2	2	2	2
16. 132	**17.** 114	**18.** 104	**19.** 124	**20.** 103
2	2	2	2	2
21. 133	**22.** 144	**23.** 242	**24.** 204	**25.** 424
2	2	2	2	2
26. 244	**27.** 414	**28.** 444	**29.** 334	**30.** 421
2	2	2	2	2

ORAL EXERCISE

Give results rapidly:

1.	**2.**	**3.**	**4.**
$2\times2+1$	$2\times2+1$	$10\times2+2$	$3\times2+2$
$4\times2+1$	$3\times2+2$	$9\times2+2$	$2\times2+2$
$5\times2+1$	$2\times2+2$	$7\times2+1$	$6\times2+1$
$6\times2+1$	$3\times2+1$	$6\times2+2$	$7\times2+1$
$4\times2+2$	$2\times8+1$	$2\times5+1$	$8\times2+2$
$3\times2+2$	$2\times3+1$	$7\times2+2$	$5\times2+2$

Carrying Figures. Multiply 26 by 2:

26	2 times 6 are 12. Write 2 and carry 1.
×2	2 times 2 are 4. 4+1=5.
52	*Write units under units and tens under tens.*

WRITTEN EXERCISE

Find these products:

1. 23 2	2. 46 2	3. 36 2	4. 96 2	5. 17 2
6. 57 2	7. 77 2	8. 212 4	9. 102 4	10. 122 5
11. 122 6	12. 212 7	13. 212 6	14. 202 8	15. 312 2
16. 226 2	17. 335 2	18. 202 5	19. 122 7	20. 102 8

21. 120×5 22. 345×2 23. 240×4 24. 125×2

Multiplying by 4.

1. Count by 4's to 48.

2. 4+4=? 2×4=? 4×2=?
4+4+4=? 3×4=? 4×3=?
4+4+4+4=? 4×4=? 4×4=?

Seven 4's=7×4. Ten 4's=10×4. Twelve 4's=12×4.

The Table of 4's. Learn this:

4×1 = 4	4×7 = 28	1×4 = 4	7×4 = 28
4×2 = 8	4×8 = 32	2×4 = 8	8×4 = 32
4×3 = 12	4×9 = 36	3×4 = 12	9×4 = 36
4×4 = 16	4×10 = 40	4×4 = 16	10×4 = 40
4×5 = 20	4×11 = 44	5×4 = 20	11×4 = 44
4×6 = 24	4×12 = 48	6×4 = 24	12×4 = 48

ORAL EXERCISE

Give answers at sight:

1. 4×6	2. 4×10	3. 4×7	4. 4×12
4×9	4×2	4×8	4×0
4×5	4×11	4×1	4×3
5. 7×4	6. 5×4	7. 8×4	8. 11×4
6×4	12×4	1×4	4×4
3×4	9×4	10×4	2×4

WRITTEN EXERCISE

Multiply:

1. 202	2. 303	3. 402	4. 112	5. 110	6. 220
4	4	4	4	4	4
7. 240	8. 151	9. 242	10. 130	11. 480	12. 182
4	4	4	4	4	4
13. 704	14. 328	15. 175	16. 283	17. 89	18. 195
4	4	4	8	6	5

19. 124	**20.** 224	**21.** 124	**22.** 424	**23.** 244	**24.** 114
3	4	5	3	7	8

25. 204	**26.** 132	**27.** 408	**28.** 444	**29.** 445	**30.** 381
4	3	4	3	2	6

Multiplying by 3.

1. Repeat the multiplication table of twos; the table of fours.

2. Count by 3 from 3 to 36.

3. $3+3+3=?$ $3\times3=?$

$3+3+3+3=?$ $4\times3=?$ $3\times4=?$

4. On your paper, complete the above table to 12×3.

The Table of 3's. Learn this:

$3\times1=3$	$3\times7=21$	$1\times3=3$	$7\times3=21$
$3\times2=6$	$3\times8=24$	$2\times3=6$	$8\times3=24$
$3\times3=9$	$3\times9=27$	$3\times3=9$	$9\times3=27$
$3\times4=12$	$3\times10=30$	$4\times3=12$	$10\times3=30$
$3\times5=15$	$3\times11=33$	$5\times3=15$	$11\times3=33$
$3\times6=18$	$3\times12=36$	$6\times3=18$	$12\times3=36$

ORAL EXERCISE

Give the answers:

1.	2.	3.	4.
3×10	10×3	3×11	7×3
3×8	5×3	3×9	3×1
$3\times7+3$	$3\times5+5$	$3\times6+6$	$8\times3+3$
$3\times4+2$	$3\times8+7$	$3\times2+4$	$4\times3+5$

WRITTEN EXERCISE

Multiply:

1. 308	2. 306	3. 434	4. 433	5. 334
3	4	5	6	7
6. 486	7. 332	8. 235	9. 489	10. 434
2	9	8	3	4
11. 243	12. 334	13. 897	14. 974	15. 344
8	6	4	2	5

ORAL EXERCISE

1. How many feet in 4 yards? In 6 yards? In 8 yards?
2. How many pints in 3 quarts? In 4 quarts?
3. How many quarts in 3 gallons? In 4 gallons?
4. How many days in 2 weeks? In 4 weeks?
5. How many weeks in 3 months? In 2 months? In 4 months?
6. How many inches in 2 feet? In 4 feet? In 3 feet?
7. How many eggs in 2 dozen? In 4 dozen? In 3 dozen?
8. How many months in 1 year? In 2 years? In 4 years? In 3 years?
9. Multiply each number in this line by 2; by 4.

   7 5 12 4 9 8 6 3

10. Multiply each number in this line by 3; by 2.

    8 6 11 5 10 9 7 4

WRITTEN EXERCISE

Multiply:

1. 265 2	**2.** 809 3	**3.** 556 4	**4.** 874 4	**5.** 166 4
6. 434 7	**7.** 334 8	**8.** 443 6	**9.** 244 5	**10.** 314 9
11. 413 4	**12.** 964 3	**13.** 738 3	**14.** 486 4	**15.** 384 3

Multiplying by 5.

$5+5=2\times5 \qquad 5+5+5+5=4\times5$

$5+5+5=3\times5 \qquad 5+5+5+5+5=5\times5$

On your paper, complete this table to 5×12.

The Table of 5's. Learn this:

$1\times5=5$	$7\times5=35$	$5\times1=5$	$5\times7=35$
$2\times5=10$	$8\times5=40$	$5\times2=10$	$5\times8=40$
$3\times5=15$	$9\times5=45$	$5\times3=15$	$5\times9=45$
$4\times5=20$	$10\times5=50$	$5\times4=20$	$5\times10=50$
$5\times5=25$	$11\times5=55$	$5\times5=25$	$5\times11=55$
$6\times5=30$	$12\times5=60$	$5\times6=30$	$5\times12=60$

ORAL EXERCISE

1.	**2.**	**3.**	**4.**
5×5	5×6	5×7	7×5
5×8	5×3	5×9	4×5
$5\times11+4$	$5\times7+5$	$4\times5+3$	$11\times5+4$
$6\times5+4$	$3\times5+2$	$10\times5+8$	$4\times5+4$

WRITTEN EXERCISE

Multiply:

1. 221 5	2. 1489 3	3. 1465 4	4. 443 6	5. 244 7
6. 243 8	7. 433 9	8. 2897 2	9. 1424 5	10. 343 6
11. 274 5	12. 1284 4	13. 737 5	14. 1186 5	15. 843 5
16. 887 4	17. 566 5	18. 1144 8	19. 444 6	20. 894 3
21. 1164 5	22. 1143 5	23. 1072 5	24. 1084 5	25. 1009 5

Multiplying by Two Figures. Multiply 38 by 21:

		PROOF
38 21 ——— 38 76 ——— 798	Here we multiply 38 by 1 and write the product with the first right-hand figure under the multiplier 1. Next we multiply 38 by 2 and write the second product 76 with the first right-hand figure 6 under the multiplier 2. Then we add the two partial products.	21 38 ——— 168 63 ——— 798

WRITTEN EXERCISE

Multiply:

1. 27 18	2. 36 19	3. 54 21	4. 43 22	5. 62 23

Find products:

6. 45	**7.** 28	**8.** 58	**9.** 67	**10.** 53
24	16	32	42	18
11. 56	**12.** 56	**13.** 74	**14.** 88	**15.** 78
32	44	23	25	33
16. 75	**17.** 39	**18.** 49	**19.** 53	**20.** 37
34	24	14	44	42
21. 46	**22.** 55	**23.** 83	**24.** 34	**25.** 92
42	23	35	33	32
26. 19	**27.** 76	**28.** 94	**29.** 45	**30.** 56
34	43	25	32	52

31. A barrel of flour weighs 196 pounds. How many pounds will there be in 5 barrels of flour?

32. If a carpenter pays 34¢ for a bundle of shingles, what will he pay for 25 bundles?

WRITTEN EXERCISE

Multiply:

1. 1336×50
2. 1242×60
3. 963×40
4. 1184×30
5. 1437×40
6. 169×53
7. 342×64
8. 178×42
9. 182×53
10. 347×24
11. 264×43
12. 275×31
13. 483×15
14. 245×54
15. 138×32
16. 149×44
17. 157×55
18. 283×31
19. 297×25
20. 341×13
21. 252×25

ORAL DRILL EXERCISE

	A	*B*	*C*	*D*
	Give answers:			
1.	6×5	5×12	½ of 10	8×5
2.	½ of 20	16÷4	27+11	⅓ of 9
3.	54+13	71+13	7×4	25÷5
4.	4×8	¼ of 12	15÷5	9×4
5.	20÷5	7×5	3×8	36+12
	Add:			
6.	2 3 23	4 5 36	5 7 45	6 5 52
7.	9 23	8 84	6 34	7 55
	Give answers:			
8.	3×5+2	24÷3	5×5+3	43−12
9.	⅓ of 30	3×3+4	⅓ of 15	6×3+2
10.	16÷4	½ of 12	45÷9	½ of 14
11.	54−13	2×5+1	36−7	27÷3
12.	7×4×3	36÷3	2×4+4	7×2+8
	Subtract:			
13.	33 81 −7 −6	54 92 −9 −8	66 38 −7 −9	75 49 −8 −5
14.	93 31 −8 −5	42 84 −4 −9	71 28 −2 −7	65 37 −6 −8

GENERAL ORAL PROBLEMS

1. There are 18 girls in Jennie's class and 11 in Martha's. How many in both?

2. John bought an orange for 3¢ and a sandwich for 5¢. How much was left of his dime?

3. Louis won 12 games of marbles and Fred won 13. How many did both win?

4. I bought a strap for 6 cents and received 19 cents in change. How much money did I give the clerk?

5. I sold for 25 cents a book costing 40 cents. How much did I lose?

6. We have 35 pupils in our class. 15 are boys. How many are girls?

7. There are 8 seats in each row. There are 4 rows. How many seats in the room?

8. My book cost 3 cents. Harry's cost 4 times as much. How much did Harry's cost?

9. John sold 8 papers, Max 7, and Tom 10. How many papers were sold in all?

10. Ely's father paid $18 for a saddle and twice as much for a tent. How much did the tent cost?

11. We have 42 paint pans. How many shall we have if we borrow 18 more?

12. Cherries cost 4 cents a pound. What will an 8-pound box cost?

13. If a man saves $9 each month, how much will he save in 5 months?

GENERAL WRITTEN PROBLEMS

1. Joe's mother spent 25 cents for fruit, 18 cents for milk, and 82 cents for meat. How much did she spend in all?

2. We have 39 pupils in our class. Last term we had 27. How many more are there this term?

3. The grocer buys sugar for 6 cents a pound and sells it for 8 cents a pound. How much will he make if he sells 134 pounds?

4. The gasman earns $14 a week. How much will he earn in 5 weeks?

5. His wife spends $2.50 a week for meat and twice as much for groceries. How much does she spend for meat and groceries together?

6. The first row had 34 spelling words correct; the second had 11 more correct. How many words did both rows have correct?

7. Harry earned 24 cents selling papers and four times as much delivering orders. How much did he earn in all?

8. Bob's father bought a bicycle for $18.45 and sold it for $23.40. How much did he gain?

9. Our spelling lesson lasted 13 minutes. Our drawing lesson will last 4 times as long. How many minutes longer will that be?

10. My sister's suit cost $12.35, her shoes $3.50, and her hat $4.75. What did they all cost?

11. Last month our gas bill was $4.18. This month it is $5.10. How much larger is it this month than last?

12. Last week we used 49 sheets of drawing paper. This week we used 17 more. Find how many sheets we used in the 2 weeks.

13. A storekeeper had $78.32 in his cash drawer in the morning. At night he had $259.25. How much had he taken in during the day?

14. A man bought a pair of shoes for $6.85. He gave the clerk a 20-dollar bill. Find how much change he received.

15. A car conductor started on a trip with $4.95. At the end of the trip he had $21.10. How much did he receive on the trip?

16. If each day we spend 14 minutes in exercising, in 4 days how many minutes do we spend in exercising?

17. The rent of our house is $52 a month. How much will our rent be in one year?

18. The baker delivers 143 loaves of bread every day to a restaurant. How many loaves does he deliver in 5 days?

19. A collector of the gas company had $11.49 in his pocket in the morning. If he collected bills amounting to $76.14, how much money had he then?

20. An automobile company sold 5 machines at $424 each. How much were they all worth?

VI. DIVISION

Dividing by 2.

1. If one orange costs 2 cents, what will five oranges cost?

2. If one orange costs 2 cents, how many can you buy for 10 cents? For 8 cents? For 20 cents? How many 2's in 20? In 4? In 10?

Instead of saying 10 contains five 2's, we may say 10 divided by 2 is 5.

Instead of the words *divided by* we may use the division sign (÷) and write: $10 \div 2 = 5$ or $2\overline{)10}$
5

In like manner we may write:

$4 \div 2 = ?$ or $2\overline{)4}$? $20 \div 2 = ?$ or $2\overline{)20}$?

3. Read these examples and give the answers:

$2\overline{)22}$	$2\overline{)24}$	$2\overline{)44}$	$2\overline{)46}$	$2\overline{)48}$	$2\overline{)40}$
$2\overline{)68}$	$2\overline{)64}$	$2\overline{)12}$	$2\overline{)28}$	$2\overline{)66}$	$2\overline{)42}$
$2\overline{)80}$	$2\overline{)82}$	$2\overline{)84}$	$2\overline{)86}$	$2\overline{)88}$	$2\overline{)62}$

In dividing these numbers by 2, we have divided them into 2 equal parts called *halves*.

To express *one half* in figures we write $\frac{1}{2}$; for example, $\frac{1}{2}$ of $10 = 5$. In reading this, we say, "One half of 10 is 5."

Factors. The number 12=6×2. The 6 and the 2 in this case are called **factors** of 12.

When one factor is 2, what is the other factor in 6? 4? 8? 10? 22? 18? 24? 30? 40? 60? 32?

ORAL EXERCISE

Supply the multipliers:

1.	2.	3.	4.
2×?= 4	2×?=16	2×?=14	2×?=24
2×?=20	2×?=30	2×?=18	2×?=10
2×?=12	2×?= 8	2×?=40	2×?=22

The Table of 2's. Learn this:

2÷2=1	14÷2= 7	½ of 2=1	½ of 14= 7
4÷2=2	16÷2= 8	½ of 4=2	½ of 16= 8
6÷2=3	18÷2= 9	½ of 6=3	½ of 18= 9
8÷2=4	20÷2=10	½ of 8=4	½ of 20=10
10÷2=5	22÷2=11	½ of 10=5	½ of 22=11
12÷2=6	24÷2=12	½ of 12=6	½ of 24=12

ORAL EXERCISE

Give the answers:

1.	2.	3.	4.	5.	6.
6÷2	12÷2	16÷2	½ of 4	½ of 2	½ of 20
4÷2	10÷2	22÷2	½ of 6	½ of 12	½ of 18
8÷2	20÷2	2÷2	½ of 10	½ of 8	½ of 16
2)14	2)24	2)22	½ of 14	½ of 24	½ of 22
2)28	2)40	2)44	½ of 42	½ of 28	½ of 44

Divisors of One Figure. Divide 68 by 2:

2)68	6÷2=3. We write the 3 under the 6.
34	8÷2=4. We write the 4 under the 8.

Proof in division: If we multiply the quotient 34 by the divisor 2, we get the dividend 68. This proves our answer to be correct.

WRITTEN EXERCISE

Divide and prove:

1. 2)40	2. 2)44	3. 2)28	4. 2)80
5. 2)88	6. 2)42	7. 2)62	8. 2)66
9. 2)82	10. 2)24	11. 2)22	12. 2)20
13. 2)60	14. 2)26	15. 2)46	16. 2)68
17. 2)248	18. 2)264	19. 2)284	20. 2)442
21. 2)220	22. 2)440	23. 2)428	24. 2)644
25. 882÷2	26. 868÷2	27. 828÷2	28. 684÷2
29. 884÷2	30. 840÷2	31. 468÷2	32. 448÷2

Terms in Division. Consider the example:

$$68 \div 2 = 34$$

The number to be divided, 68 in this example, is called the **dividend.**

The number by which we divide, 2 in this example, is the **divisor.**

The result in division is called the **quotient.**

Dividing by 4.

1. If one ruler costs 4 cents, what will 6 rulers cost?

2. If one ruler costs 4 cents, how many can you buy for 24 cents? For 8 cents? For 20 cents? For 12 cents? How many can you buy for 40 cents?

How many 4's in 24? In 8? In 12? In 20? In 40?

$24 \div 4 = ?$ $8 \div 4 = ?$ $12 \div 4 = ?$ $20 \div 4 = ?$ $40 \div 4 = ?$

3. Divide the following:

4)8	4)20	4)16	4)40	4)28	4)84
4)32	4)12	4)24	4)4	4)44	4)88

In dividing these numbers by 4, we have divided them into 4 equal parts.

We call these equal parts *fourths* or *quarters*. One fourth is written $\frac{1}{4}$.

Draw $\frac{1}{4}$ of a circle; $\frac{1}{4}$ of a pie; fold a piece of paper into quarters.

The Table of 4's. Learn this:

$4 \div 4 = 1$	$28 \div 4 = 7$	$\frac{1}{4}$ of $4 = 1$	$\frac{1}{4}$ of $28 = 7$
$8 \div 4 = 2$	$32 \div 4 = 8$	$\frac{1}{4}$ of $8 = 2$	$\frac{1}{4}$ of $32 = 8$
$12 \div 4 = 3$	$36 \div 4 = 9$	$\frac{1}{4}$ of $12 = 3$	$\frac{1}{4}$ of $36 = 9$
$16 \div 4 = 4$	$40 \div 4 = 10$	$\frac{1}{4}$ of $16 = 4$	$\frac{1}{4}$ of $40 = 10$
$20 \div 4 = 5$	$44 \div 4 = 11$	$\frac{1}{4}$ of $20 = 5$	$\frac{1}{4}$ of $44 = 11$
$24 \div 4 = 6$	$48 \div 4 = 12$	$\frac{1}{4}$ of $24 = 6$	$\frac{1}{4}$ of $48 = 12$

WRITTEN EXERCISE

Divide:

1. 4)448	2. 2)828	3. 4)808	4. 2)428
5. 4)840	6. 2)408	7. 4)888	8. 2)844
9. 4)880	10. 2)248	11. 4)480	12. 2)486
13. 4)848	14. 2)668	15. 4)484	16. 2)640
17. 248÷2	18. 662÷2	19. 848÷4	20. 282÷2

Finding Factors

Give the missing factors:

1.	2.	3.	4.
22=?×11	32=?×4	9=?×3	44=?×4
33=?× 3	24=?×4	10=?×5	18=?×2
20=?× 2	12=?×4	8=?×4	40=?×4
28=?× 4	6=?×2	27=?×3	36=?×4
48=?× 4	16=?×4	21=?×3	12=?×2

5. If one factor is 3, what is the other factor in 9? 21? 27? 12? 15? 24? 30? 36? 33? 3?

6. If one factor is 4, what is the other factor in 36? 24? 44? 4? 12? 16? 28? 40? 32? 20?

Dividing by 3.

1. How many 3's in 6? 6÷3=? $\frac{1}{3}$ of 6=? 3)6 ?

2. How many 3's in 9? 9÷3=? $\frac{1}{3}$ of 9=? 3)9 ?

3. How many 3's in 12? 12÷3=? $\frac{1}{3}$ of 12=? 3)12 ?

The Table of 3's. Learn this:

3÷3=1	21÷3= 7	$\frac{1}{3}$ of 3=1	$\frac{1}{3}$ of 21= 7
6÷3=2	24÷3= 8	$\frac{1}{3}$ of 6=2	$\frac{1}{3}$ of 24= 8
9÷3=3	27÷3= 9	$\frac{1}{3}$ of 9=3	$\frac{1}{3}$ of 27= 9
12÷3=4	30÷3=10	$\frac{1}{3}$ of 12=4	$\frac{1}{3}$ of 30=10
15÷3=5	33÷3=11	$\frac{1}{3}$ of 15=5	$\frac{1}{3}$ of 33=11
18÷3=6	36÷3=12	$\frac{1}{3}$ of 18=6	$\frac{1}{3}$ of 36=12

ORAL EXERCISE

Divide:

1. $\frac{1}{3}$ of 6　$\frac{1}{3}$ of 3　$\frac{1}{3}$ of 12　$\frac{1}{3}$ of 18　$\frac{1}{3}$ of 15　$\frac{1}{3}$ of 9
   $\frac{1}{3}$ of 21　$\frac{1}{3}$ of 24　$\frac{1}{3}$ of 33　$\frac{1}{3}$ of 27　$\frac{1}{3}$ of 36　$\frac{1}{3}$ of 30

2. 3)33　3)24　3)27　3)15　3)3　3)6
   3)12　3)30　3)18　3)9　3)21　3)36

Supply the missing numbers:

3.	4.	5.	6.	7.
? ÷3=5	? ÷60=3	? ÷4=11	22÷?=11	28÷?=7
18÷3=?	33÷ ? =3	24÷4= ?	? ÷7=2	? ÷2=5
36÷?=3	? ÷3 =9	48÷?=12	20÷?=2	? ÷4=9

WRITTEN EXERCISE

Divide:

1. 3)337	2. 2)464	3. 4)488	4. 3)390
5. 4)4088	6. 2)8066	7. 3)3036	8. 3)363
9. 4)888	10. 2)246	11. 3)939	12. 2)2064
13. 4)4848	14. 3)3636	15. 4)8048	16. 3)993

17. 2)468	**18.** 4)804	**19.** 3)633	**20.** 4)8844
21. 2)6688	**22.** 3)3090	**23.** 3)3933	**24.** 3)3006
25. 2)2484	**26.** 4)484	**27.** 3)369	**28.** 2)6048

Dividing by 5.

1. If 1 pint of oil costs 5 cents, how many pints can you buy with 40 cents? How many 5's in 40? What is $40 \div 5$? Find $\frac{1}{5}$ of 40. 5)40 ?

2. I want to give out 25 crayons to 5 boys. How many will each boy receive? How many 5's in 25? What is $\frac{1}{5}$ of 25? $25 \div 5$? 5)25. ?

3. If there are 30 days in a month, how many days in $\frac{1}{2}$ the month? $\frac{1}{3}$ the month? $\frac{1}{5}$ the month?

The Table of 5's. Learn this:

$5 \div 5 = 1$	$35 \div 5 = 7$	$\frac{1}{5}$ of $5 = 1$	$\frac{1}{5}$ of $35 = 7$
$10 \div 5 = 2$	$40 \div 5 = 8$	$\frac{1}{5}$ of $10 = 2$	$\frac{1}{5}$ of $40 = 8$
$15 \div 5 = 3$	$45 \div 5 = 9$	$\frac{1}{5}$ of $15 = 3$	$\frac{1}{5}$ of $45 = 9$
$20 \div 5 = 4$	$50 \div 5 = 10$	$\frac{1}{5}$ of $20 = 4$	$\frac{1}{5}$ of $50 = 10$
$25 \div 5 = 5$	$55 \div 5 = 11$	$\frac{1}{5}$ of $25 = 5$	$\frac{1}{5}$ of $55 = 11$
$30 \div 5 = 6$	$60 \div 5 = 12$	$\frac{1}{5}$ of $30 = 6$	$\frac{1}{5}$ of $60 = 12$

ORAL EXERCISE

1. Bread is 5 cents a loaf. How many loaves can I buy for 15¢? 40¢? 30¢? 25¢? 60¢? 55¢? 35¢? 20¢?

Give the missing numbers:

2.	3.	4.	5.
15÷?=3	?÷5=4	?÷4=7	?÷5=11
60÷?=12	?÷5=9	?÷3=6	55÷5=?
25÷5=?	?÷5=5	?÷5=6	30÷5=?
40÷?=8	?÷5=7	?÷2=6	45÷5=?
5÷?=1	10÷?=2	?÷5=1	24÷?=6

6. If one factor is 5, what is the other factor in 20? 30? 40? 50? 65?

7. If one factor is 3, what is the other factor in 18? 30? 24? 15? 27? 36?

WRITTEN EXERCISE

Divide:

1. 5)55	2. 5)550	3. 5)5055	4. 5)30
5. 5)305	6. 5)3050	7. 5)3555	8. 5)10
9. 5)105	10. 5)1050	11. 5)50	12. 5)500
13. 5)5005	14. 5)25	15. 5)2505	16. 5)4050

WRITTEN REVIEW

Find the quotients:

1. 5)45	2. 5)4500	3. 5)5555	4. 3)1503
5. 4)1200	6. 3)3303	7. 4)1248	8. 4)2884
9. 4)3208	10. 3)150	11. 3)1557	12. 3)2493
13. 5)2530	14. 3)2709	15. 4)1604	16. 5)2005
17. 4)2420	18. 3)1809	19. 5)3550	20. 5)5050

ORAL PROBLEMS

Mailing Letters.

1. I bought 14 two-cent stamps. How much did I pay for them?

2. Helen bought 3 five-cent stamps and 2 postal cards. How much did she pay?

3. Ely needs 4 stamps. His father will need 12 times as many. How many will his father need?

4. George has a dollar. How many postal cards can he buy? How many two-cent stamps? How many five-cent stamps?

5. Will mailed 5 letters. Harry mailed 5 times as many. How many did Harry mail? How many did they both mail?

6. The postman found 44 letters in the first box and $\frac{1}{4}$ as many in the second. How many were in the second?

WRITTEN PROBLEMS

1. If I buy 13 two-cent stamps, 11 postal cards and 3 three-cent stamps, how much change shall I receive from a one-dollar bill?

2. I want to send 22 letters to Germany. I must place a five-cent stamp on each. How much will the stamps cost?

3. With 78 cents, how many letters can I mail to Albany, if each one requires a two-cent stamp?

4. I want to buy 29 postal cards, 3 five-cent stamps, and 19 two-cent stamps. What will they cost?

ORAL DRILL EXERCISE

	A	*B*	*C*	*D*
	Divide:			
1.	40÷2	27÷3	40÷5	36÷4
2.	5)45	4)44	5)30	2)18
3.	4)12	5)35	2)22	3)33
4.	16÷2	24÷3	60÷5	48÷4
5.	4)28	5)50	3)36	5)45
6.	40 80 ×2 ×3	50 90 ×3 ×4	70 20 ×5 ×4	60 80 ×3 ×5
	Find:			
7.	$\frac{1}{2}$ of 40	$\frac{1}{4}$ of 40	$\frac{1}{3}$ of 60	$\frac{1}{5}$ of 35
8.	$\frac{1}{5}$ of 60	$\frac{1}{5}$ of 45	$\frac{1}{2}$ of 24	$\frac{1}{4}$ of 32
9.	46+8	40÷4	62+7	$\frac{1}{3}$ of 21
10.	8×5	33−11	45÷5	71+12
11.	25÷5	$\frac{1}{2}$ of 16	6×5+11	8×4+2
12.	$\frac{1}{3}$ of 24	4×3 +5	$\frac{1}{4}$ of 12	3×8+2
13.	28−9	$\frac{1}{4}$ of 24	46−20	54−17
14.	50 20 +34 +48	40 60 +45 +29	80 50 +18 +37	70 40 +24 +56
15.	28 90 −30 −40	37 81 −20 −40	46 72 −30 −50	54 36 −20 −30
16.	3×7+9	5×12+7	7×3+5	9×4+3
17.	4×8+8	6× 4+6	8×5+4	10×3+2

GENERAL ORAL PROBLEMS

1. Van has 5 examples correct. Gertrude has 4 times as many right. How many has Gertrude correct?

2. Ely paid $4 for a hat, and 8 times as much for a coat. Find the cost of the coat.

3. He paid $\frac{1}{2}$ as much for his shoes as for his hat. Find the cost of his shoes.

4. There are 12 windows in the assembly room. $\frac{1}{2}$ of them are open. How many are open?

5. A baseball costs 12¢. What will 5 baseballs cost?

6. Mae bought 9 oranges at 4 cents each. How much did she pay for them?

7. John earned 22¢ by selling papers, and 12¢ by running errands. How much did he earn in all?

8. Tom's brother is 12 years old. He is four times as old as my brother. How old is my brother?

9. We have 47 paint pans. I have washed 20. How many have I still to wash?

10. I need 14 two-cent stamps for my father's letters. In order to buy them, how much money shall I need?

11. If my father saves $12 each month, how much will he save in 4 months?

12. How much shall I have to pay for $\frac{1}{4}$ of a pound of candy that costs 64¢ a pound?

13. If four pounds of sugar cost 28 cents, what will one pound cost?

GENERAL WRITTEN PROBLEMS

1. Oranges cost 5 cents each. What will 5 doz. cost?

2. I have 84 cents to buy two-cent stamps. How many stamps can I get?

3. How many pints of oil at 8¢ a pint can I buy for $8.80?

4. Joe's mother paid $6.86 to the butcher, $4.50 to the baker, and $1.65 to the iceman. How much did she spend?

5. Ely bought a kite for $.85 and sold it to Van for $1.25. How much money did Ely gain?

6. Find the cost of 8 pounds of steak at $.28 a pound.

7. Ely earned 11 cents yesterday and 16 cents to-day. How much more must he earn to buy a fifty-cent ball?

8. Tom sold 96 papers last week. $\frac{1}{3}$ of them were morning papers. How many were morning papers?

9. If our arithmetics cost $.44 each, what is the cost of 40 of them?

10. Our baker's bill this month is $8.40. Last month it was only $\frac{1}{2}$ as much. How much was it last month?

11. Coffee costs 28¢ a pound. We use 4 pounds a month. How much does our coffee cost every month?

12. Vinegar sells for 12¢ a quart. Find the cost of 8 gallons.

VII. FRACTIONS

Halves. This block has been divided into two equal parts called *halves*. The whole block is made up of two halves, $\frac{2}{2}$, and each part is called *one half*, $\frac{1}{2}$.

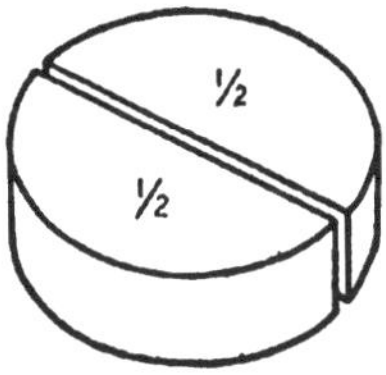

1. How would you write one and one half? Two and one half?

2. How many half blocks would you have in $3\frac{1}{2}$ blocks?

3. What is $\frac{1}{2}$ of 8? 12? 20? 4? 24? 28? 30? 36? 40? 44?

WRITTEN EXERCISE

1. There were 52 boys on the line. $\frac{1}{2}$ went to their seats. How many went to their seats?

2. Ethel had 48¢. She spent $\frac{1}{2}$ of it. How much has she left?

3. We had 76 pieces of chalk. We sent $\frac{1}{2}$ to another room. How many have we now?

4. One line is 38 inches long. Another is $\frac{1}{2}$ as long. How long are both together?

5. There are 18 boys and 8 girls on a line. This number is just $\frac{1}{2}$ of our class. How many pupils in the class?

Fourths.

1. $16 \div 4 = ?$ What is $\frac{1}{4}$ of 16?

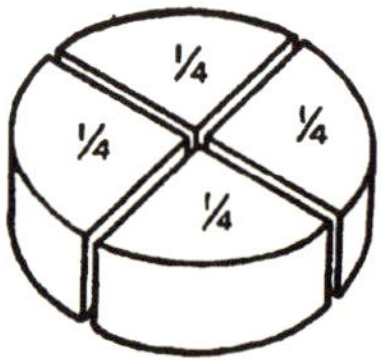

We get *one fourth* of anything by dividing by 4. This block has been divided into four equal parts. We call each part *one fourth* or *one quarter*. How many fourths in the whole block?

2. How many fourths would there be in two blocks? In 4? In $1\frac{1}{2}$? How many quarters in $\frac{1}{2}$?

3. If I divide 16 cents equally among 4 boys, how many cents will each boy receive? $16 \div 4 = ?$ $\frac{1}{4}$ of $16 = ?$ $4\overline{)16}$.
?

Equal parts of a unit, like *halves* and *fourths*, are called **fractions.**

4. Find $\frac{1}{4}$ of 8; 20; 24; 44; 32; 28; 40; 48.

WRITTEN EXERCISE

1. Harry had 36¢. If he spent $\frac{1}{4}$ of that amount for a strap, how many cents did he spend?

2. We have 20 electric lights in our room. The teacher turned on $\frac{1}{4}$ of them. How many of them were not lighted ?

3. The grocer had 60 gallons of vinegar. He sold $\frac{1}{4}$ of them. How many did he sell?

4. Yesterday he bought 32 boxes of berries. $\frac{1}{4}$ of them were too soft. How many of them were good?

Eighths. This block is divided into 8 equal parts. Each part is called *one eighth* or $\frac{1}{8}$.

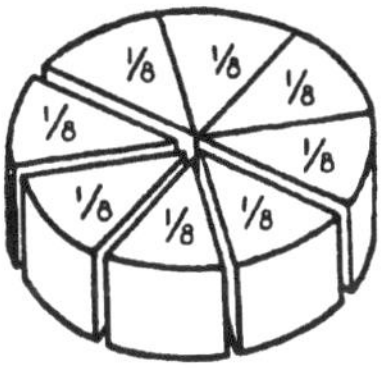

1. How many eighths in $\frac{1}{2}$ of the block? In $\frac{1}{4}$? In 2 blocks? In 3 blocks? In $1\frac{1}{8}$ blocks?

2. If I divide 24 pieces of paper among 8 pupils, how many sheets will each one receive? $24 \div 8 = ?$

$\frac{1}{8}$ of $24 = ?$ $\quad 8\overline{)24.}$?

3. How many halves in this block? How many quarters in it?

4. How many eighths in a dollar? A pie? A foot? An orange?

5. Find $\frac{1}{8}$ of 16; 8; 24; 40; 32; 48; 64; 56; 72; 80; 96.

WRITTEN EXERCISE

1. Gertrude has 16¢. If she spends $\frac{1}{8}$ of her money, how much will she spend? How much will she have left?

2. John has 24 sheets of drawing paper. If he gives $\frac{1}{8}$ of them to a row, how many sheets will each row have?

3. There are 40 pieces of chalk in the box. $\frac{1}{8}$ of them are yellow. How many pieces are yellow?

4. We have 48 pupils in our class. $\frac{1}{8}$ of them are on the honor roll. How many are on the honor roll?

Thirds.

1. How would you find *one third* of a block? How many parts would you have? How would you find $\frac{1}{3}$ of 6 marbles? How many marbles would there be in each part?

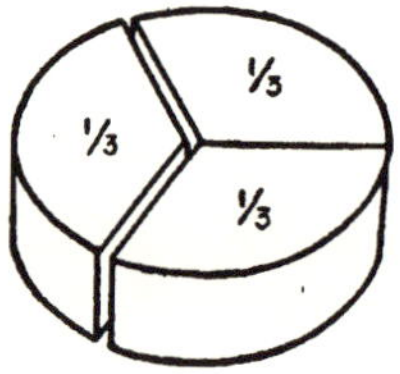

2. How would you find $\frac{1}{3}$ of 24 marbles?

$24 \div 3 = 8$. $\frac{1}{3}$ of $24 = 8$. $3\overline{)24}$ 8

3. What is $\frac{1}{3}$ of 9 boys? $\frac{1}{3}$ of 1 yard? $\frac{1}{3}$ of 15 pencils?

4. Find $\frac{1}{3}$ of 12; 3; 6; 18; 27; 21; 30; 36; 33.

WRITTEN EXERCISE

1. A line is 9 inches long. If you mark off $\frac{1}{3}$ of it, how many inches will be left?

2. Van has 21 picture cards. If he gives away $\frac{1}{3}$ of them, how many will he have left?

3. There were 27 books on one shelf. We took down $\frac{1}{3}$ of them. How many did we take down?

4. If I can buy 12 eggs for 36¢, how much shall I have to pay for $\frac{1}{3}$ of 12?

5. 3 of the electric lights in our room are not lighted. This number is just $\frac{1}{3}$ of all the lights in the room. How many lights have we?

6. Helen bought 4 postal cards. This took $\frac{1}{3}$ of her money. How much money did she have?

Fifths and Tenths.

1. How many parts in the first figure? How many in the second? What would you call each part in the first figure? What name would you give to each part in the second figure?

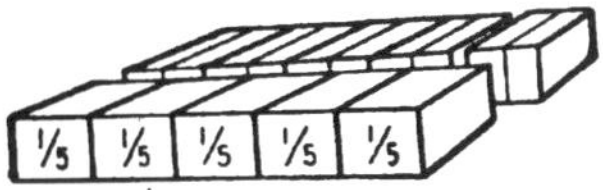

2. How many *tenths* in $\frac{1}{5}$? In $\frac{2}{5}$? How many *tenths* in $\frac{1}{2}$? In 1? In $1\frac{1}{10}$? In $1\frac{1}{2}$?

3. How many *fifths* in 1? In $1\frac{2}{5}$? In 2? In $2\frac{3}{5}$?

4. If 15 cents is divided equally among 5 boys, how much will each boy have?

$15 \div 5 = ?$ $\frac{1}{5}$ of 15? $3\overline{)15}$?

5. If 30 cents is divided equally among 10 boys, how much will each boy have?

$30 \div 10 = ?$ $\frac{1}{10}$ of $30 = ?$ $10\overline{)30}$?

6. Find $\frac{1}{5}$ of 10; 50; 25; 35; 40; 55. Find $\frac{1}{10}$ of 10; 30; 70; 100; 90; 40.

WRITTEN EXERCISE

1. We have 45 pupils in one class. $\frac{1}{5}$ of them were not promoted. How many were not promoted?

2. There are 30 words on the board. If we write them 5 in a row, how many rows will there be?

3. Forty books are to be divided equally into 5 piles. How many will there be in each pile? If there were eighty books, how many would there be in each pile?

4. There are thirty days in a month. $\frac{1}{10}$ of them have gone. How many days of the month have gone?

5. Harry had a dollar. He spent $\frac{1}{10}$ of it for carfare. How much has he left?

REVIEW OF FRACTIONS

1. If we divide a number into 2 equal parts, each part is called *one half* or $\frac{1}{2}$. Find $\frac{1}{2}$ of 44; 36; 18; 66; 80; 100; 50; 24.

2. If we divide a number into 3 equal parts, each part is called *one third* or $\frac{1}{3}$. What is $\frac{1}{3}$ of 24? 36? 60? 90? 15? 30? 45? 48?

3. If we divide a number into 4 equal parts, each part is called *one fourth* or $\frac{1}{4}$. Find $\frac{1}{4}$ of 16; 40; 88; 32; 4; 36; 48; 60.

4. If we divide a number into 5 equal parts, each part is called *one fifth* or $\frac{1}{5}$. Find $\frac{1}{5}$ of 15; 60; 45; 25; 30; 55; 5; 35.

5. If we divide a number into 8 equal parts, each part is called *one eighth* or $\frac{1}{8}$. What is $\frac{1}{8}$ of 16? 24? 40? 32? 8?

6. If we divide a number into 10 equal parts, each part is called *one tenth* or $\frac{1}{10}$. What is $\frac{1}{10}$ of 10? 30? 50? 90? 70? 60? 40? 20?

ORAL DRILL EXERCISE

	A	*B*	*C*	*D*
1.	5×5	$\frac{1}{4}$ of 32	3)33	5)? 6
2.	?×8=40	8×5	$\frac{1}{3}$ of 27	32÷?=8
3.	$\frac{1}{7}$ of 35	5)40	4×9	4)28
4.	4)44	$\frac{1}{5}$ of 35	?÷10=3	3×11
5.	24÷3=	2)? 12	4)? 4	$\frac{1}{2}$ of 30

Find:

6.	$\frac{1}{2}$ of 30	$\frac{1}{4}$ of 44	$\frac{1}{5}$ of 25	$\frac{1}{3}$ of 18
7.	$\frac{1}{2}$ of 60	$\frac{1}{4}$ of 24	$\frac{1}{5}$ of 40	$\frac{1}{3}$ of 27
8.	$\frac{1}{3}$ of 36	$\frac{1}{5}$ of 35	$\frac{1}{4}$ of 16	$\frac{1}{2}$ of 14
9.	$\frac{1}{3}$ of 18	$\frac{1}{5}$ of 15	$\frac{1}{4}$ of 24	$\frac{1}{2}$ of 32
10.	$\frac{1}{2}$ of 44	$\frac{1}{3}$ of 24	$\frac{1}{5}$ of 30	$\frac{1}{4}$ of 16

Give answers:

11.	32÷?=4	?×15=30	$\frac{1}{4}$ of 44	?×7=35
12.	?×8=32	18÷?=9	7×?=28	?)27 9
13.	$\frac{1}{2}$ of 14	?×3=24	30÷?=15	?×3=15
14.	?)60 5	$\frac{1}{3}$ of 33	?×6=30	25÷?=5
15.	?×9=45	?)32 8	?)40 5	$\frac{1}{5}$ of 30

Add:

16.	7+5+4	9+5+7	3+2+6	2+7+5
17.	9+3+2	7+9+4	9+8+5	5+6+4
18.	8+7+5	6+3+4	6+7+3	3+5+7

GENERAL ORAL PROBLEMS

1. Louis has $.18 left out of his half-dollar. How much did he spend?

2. Margaret's sister bought a veil for $.32. What would 4 veils cost at this price?

3. We have 44 pupils in one class. $\frac{1}{4}$ of them have been late once this term. How many have been late once?

4. I bought a book for 36¢ and sold it for 11 cents more. How much did I get for it?

5. How many 2's in 10? If apples are 2 cents each, how many can I buy for 10 cents?

6. If sugar is 8¢ a pound, how many pounds can I get for 24 cents?

7. If tea is sold at 44¢ a pound, what will a half pound cost?

8. There are 11 girls in one row. This number is just $\frac{1}{4}$ of the class. How many in the class?

9. Ethel has 36¢. How many dozen peaches can she buy at 12 cents a dozen?

10. Harry saves $2 a week. He has saved $16. How many weeks did it take to save this amount?

11. A grocer had 95 gallons of oil. How many had he after selling 69 gallons?

12. If oranges cost 5 cents each, find the cost of 4 dozen.

13. Van has 80 stamps in his stamp album. $\frac{1}{5}$ of them are French. How many are French?

GENERAL WRITTEN PROBLEMS

1. Find the cost of 82 pairs of skates at $5 a pair.

2. I want to divide $65 equally among 5 people. How much shall I give to each?

3. A fruit stand has 75 apples. This is 5 times as many apples as the stand on the next block has. How many apples are there on the stand on the next block?

4. I bought 6 yards of ribbon. How many pieces, each 2 feet in length, can I get out of it?

5. Gertrude used 18 eggs in the kitchen last week. Last month she used 6 times as many. How many did she use last month?

6. If I arrange our 45 rulers in groups of 5 each, how many groups shall I have?

7. If I send 27 rulers to Miss S., how many shall we have left?

8. $\frac{1}{3}$ of our 75 books are being used. How many are being used?

9. Van has $16.24; John has $22.88. How much more has John than Van? How much have both?

10. At $32 a dozen, find the cost of 9 dozen books.

11. If 6 books cost $14, how much would 18 cost at the same rate?

12. Harry has 18 German stamps and 4 times as many United States stamps. How many has he in all?

13. On one floor of a school there are 455 pupils. $\frac{1}{5}$ of them are at assembly. How many are at assembly?

14. A man had $7214 in the bank. He drew out $3927. How much has he in the bank now?

15. A delivery automobile is worth $2240. Find how much 4 of them are worth.

16. A piano cost $345, a sofa $48.25, a table $17.75, and a rug $13.20. How much did all cost?

17. A storekeeper's sales amounted to $891.03 on Thursday, $716.22 on Friday, and $1023.78 on Saturday. Find the amount of the sales for the three days.

18. A man sold an automobile for $2350. If he paid $1980 for it, how much did he gain?

19. Mr. Henry had $211.26 in a bank. Yesterday he put $1476.22 in the bank. How much has he there now?

20. A grocer has 335 ounces of pepper. If he puts 5 ounces in a bag, how many bags can he fill?

21. A farmer has 484 pear trees. If he cuts down every fourth tree, how many will he cut down?

22. The distance between two stations is 8440 feet. If there are 5280 feet in a mile, how much more than a mile is the distance between these stations?

23. The rent of an apartment is $47 a month. Find the rent for 4 months.

24. Mr. Hickey bought a rug for $12.35, a couch for $25.75, 3 chairs for $15, and a sofa for $12.30. How much did he spend on these articles?

VIII. MEASURES

One Dozen. Twelve pencils are called a *dozen* pencils. 12 peaches are called a *dozen* peaches.

We may write dozen **doz.**

1. Learn the following:

12 things are one dozen (doz.)

2. How many eggs in one dozen? In two dozen? In three dozen? How many apples in $\frac{1}{4}$ doz.? In $\frac{1}{3}$ doz.? In $\frac{1}{2}$ doz.?

3. If 1 dozen apples cost 20 cents, what will 2 dozen cost? $\frac{1}{2}$ doz.? $1\frac{1}{2}$ doz.? How many bananas in 4 dozen?

4. How many books in 3 dozen? In $2\frac{1}{2}$ doz.? In 4 doz.?

Pounds and Ounces.

1. How can you find out how much a book weighs?

2. Learn the following:

16 ounces (oz.) = 1 pound (lb.)

If our book weighs 1 pound, then there are 16 ounces in it.

3. How many ounces in 2 pounds of sugar? In 3 lb.? In $\frac{1}{2}$ lb.? In $\frac{1}{4}$ lb.?

4. What is $\frac{1}{2}$ of 16 oz.? $\frac{1}{4}$ of 16 oz.? $\frac{1}{8}$ of 16 oz.?

Inch, Foot, Yard.

1. How many feet in 2 yards? How many in $\frac{1}{3}$ yard? How many inches in $\frac{1}{2}$ foot? In $\frac{1}{4}$ foot? In $1\frac{1}{3}$ ft.?

2. Learn the following:

12 inches (in.)	= 1 foot (ft.)
3 feet (ft.)	= 1 yard (yd.)

3. What means would you use to find the length of this page? Of your desk? The blackboard? This room?

4. How many inches long is your desk? How many feet long is the teacher's desk?

5. What is the length in feet of the blackboard? Measure the length of the room.

6. How much longer is 1 foot than 1 inch?

ORAL EXERCISE

Find:

1.	2.	3.	4.
$\frac{1}{2}$ of 6 inches	$\frac{1}{3}$ of 9 yd.	$\frac{1}{4}$ of 12 yd.	$\frac{1}{5}$ of 20 in.
$\frac{1}{2}$ of 10 feet	$\frac{1}{3}$ of 15 ft.	$\frac{1}{4}$ of 16 in.	$\frac{1}{8}$ of 16 ft.
$\frac{1}{2}$ of 8 yards	$\frac{1}{3}$ of 30 in.	$\frac{1}{4}$ of 40 ft.	$\frac{1}{10}$ of 30 yd.

Change:

5.	6.	7.
2 ft. to in.	3 yd. to ft.	24 in. to feet
$1\frac{1}{2}$ ft. to in.	$1\frac{1}{3}$ yd. to ft.	9 ft. to yd.
$\frac{1}{4}$ ft. to in.	$\frac{1}{3}$ ft. to in.	18 in. to feet

Pint, Quart, Gallon.

1. How many pints in 2 quarts of milk? How many quarts in 2 gallons of ice cream?

2. Learn this:

2 pints (pt.)	= 1 quart (qt.)
4 quarts (qt.)	= 1 gallon (gal.)

3. How many quarts in $\frac{1}{2}$ gallon? In $\frac{1}{4}$ gal.? In 3 gal.? In 5 gal.?

4. How many times do you have to fill a pint measure to make one quart? To make 1 gallon? To make 2 gal.?

5. What part of 1 quart is 1 pint? What part of 1 gal. is 1 pint?

ORAL EXERCISE

Find:

1.	2.	3.
$\frac{1}{2}$ of 6 qt.	$\frac{1}{4}$ of 16 gal.	$\frac{1}{8}$ of 8 gal.
$\frac{1}{2}$ of 12 pt.	$\frac{1}{4}$ of 20 qt.	$\frac{1}{8}$ of 24 pt.
$\frac{1}{3}$ of 9 gal.	$\frac{1}{5}$ of 10 qt.	$\frac{1}{10}$ of 20 pt.
$\frac{1}{3}$ of 21 pt.	$\frac{1}{5}$ of 25 gal.	$\frac{1}{10}$ of 50 gal.

Telling the Time.

1. Where is the minute hand at 9 o'clock? Where is the hour hand? Where are the hands at 12 o'clock? At 3 o'clock?

2. How many 5-minute spaces does the hour hand pass over between 9 o'clock and 12 o'clock? Between 12 o'clock and 3 o'clock?

3. Read all the figures on the clock.

4. It takes the minute hand 5 minutes to pass over 1 space. How many minutes will it take it to pass over 12 spaces or 1 hour?

5. Learn this:

60 minutes (min.) = 1 hour (hr.)

6. How long will it take the minute hand to pass from XII to VI? To III? To IX? To V? To VIII?

7. When the minute hand is at XII, what time is it if the hour hand is at III? At VI? At X? At II?

8. Move the minute hand to VI and the hour hand half way between X and XI. What time does the clock tell now?

9. Read the time on the school clock? What time will it be in 10 minutes. In 20 minutes? In 30 minutes? In 60 minutes?

ORAL EXERCISE

Tell the time when:

The minute hand is:	The hour hand is:
1. At VI	Between XII and I
2. At III	A little past XII
3. At VI	At X
4. At IX	Between XI and XII
5. At III	A little past XII
6. At IX	Between XII and I
7. At IV	A little past II

Time Measure.

1. Learn this table:

60 seconds (sec.)	= 1 minute (min.)
60 minutes	= 1 hour (hr.)
24 hours	= 1 day (da.)
7 days	= 1 week (wk.)
4 weeks (about)	= 1 month (mo.)
12 months	= 1 year (yr.)

2. How many seconds in ½ a minute? How many minutes in ½ an hour? In ¼ hour? In 1½ hours?

3. How many hours in 2 days? In ½ day? How many days in 2 weeks? In 3 weeks? In 4 weeks? How many months in 2 years? In 3 years? In 4 years? In 5 years?

4. If a clock strikes twice every hour, how many times will it strike in 12 hours? In one day?

5. How many months in ½ year? In ¼ yr.?

ORAL DRILL EXERCISE

	A	*B*	*C*	*D*
	Give answers:			
1.	?÷4=12	?÷5=12	?÷3=12	?÷2=11
2.	4)48	3)36	3)34	2)24
3.	5)55	5)20	4)28	5)45
4.	$\frac{1}{5}$ of 35	$\frac{1}{10}$ of 70	$\frac{1}{4}$ of 48	$\frac{1}{10}$ of 60
5.	$\frac{1}{8}$ of 24	$\frac{1}{5}$ of 35	$\frac{1}{5}$ of 55	$\frac{1}{8}$ of 40

Subtract:

6.	54	37	63	26	72	47	81	38
	18	14	19	15	17	16	19	16
7.	32	28	43	33	67	57	68	54
	17	15	18	11	12	15	14	18

Multiply and add as indicated:

8.	8×3+7	4×6+8	5×8+7	3×12+10
9.	9×4+4	6×4+2	7×2+6	5×11+ 9
10.	3×8+5	8×5+3	8×5+4	8× 5+ 7
11.	5×4+7	4×3+8	4×3+5	9× 2+ 8
12.	3×11+7	3×12+3	2×12+[illegible]	4×12+ 2

Find:

13.	$\frac{1}{2}$ of 12	$\frac{1}{5}$ of 25	$\frac{1}{5}$ of 60	$\frac{1}{2}$ of 60
14.	$\frac{1}{4}$ of 24	$\frac{1}{4}$ of 32	$\frac{1}{4}$ of 28	$\frac{1}{2}$ of 18
15.	$\frac{1}{3}$ of 36	$\frac{1}{5}$ of 35	$\frac{1}{4}$ of 24	$\frac{1}{5}$ of 40
16.	$\frac{1}{5}$ of 45	$\frac{1}{4}$ of 48	$\frac{1}{3}$ of 30	$\frac{1}{3}$ of 36
17.	$\frac{1}{2}$ of 50	$\frac{1}{3}$ of 33	$\frac{1}{5}$ of 45	$\frac{1}{4}$ of 24

GENERAL ORAL PROBLEMS

1. Nine years ago Helen was 7 years old. How old is she now?

2. A box contains 48 pieces of chalk. $\frac{1}{4}$ of them are red. How many are white?

3. Sugar is 8 cents a pound. How much change shall I get from a quarter if I buy 2 pounds?

4. At 3 cents each, how many pencils can I buy for 15 cents?

5. Mary paid $2 to the grocer and $2.25 to the butcher. How much did she pay to both?

6. The milkman had 32 quarts of milk. He ordered 12 qts. more. How much had he then?

7. He collected 48 bottles. 12 were broken. How many were unbroken?

8. He left us 4 quarts of milk every day for 11 days. How many quarts did we buy in all?

9. Van has 18 cents with which to buy stamps. How many two-cent stamps can he get?

10. How many loaves of bread at 5 cents each can Gertrude buy for 25 cents? At the same rate, how many can she buy for 75 cents?

11. Harry bought a pencil for 4 cents. He paid 3 times as much for a pencil box. How much did the pencil box cost?

12. Oil costs 40 cents a gallon. What will 2 quarts cost? Two pints?

GENERAL WRITTEN PROBLEMS

1. Ely saved 7 cents every day from June 2 to June 18. How much did he save?

2. Harry won 24 games of marbles. We won $\frac{1}{4}$ as many. How many did we win?

3. Harry won 72 marbles. If he wins 19 marbles more, how many will he have?

4. Pads cost 3 cents each. How many can I buy for 69 cents?

5. Gertrude gave the iceman 65 cents, the butcher $3.14, and the gasman $1.88. How much did she spend?

6. I bought $\frac{1}{4}$ of a yard of ribbon at 48¢ a yard. What did I pay for it?

7. Helen paid $.12 for one quarter of a pouna of tea. What would 2 pounds cost at this price?

8. A card was 72 inches long. It was divided into 8 equal parts. How many inches were there in each piece ?

9. There are 94 pages in a spelling book and 78 pages in a reader. How many pages in both books?

10. A man had 48 roses which he makes in bunches of 4 roses each. How many bunches can he make?

11. There are 931 bricks in a pile, and a workman uses 488 for a pavement. How many bricks are left over?

12. There are 2 dozen cakes of soap in one box. How many cakes are there in 32 boxes?

13. A dealer ships 6 pairs of roller skates in a box. How many boxes will he need to ship 186 pairs of skates?

14. In his store he has 72 sweaters, 112 worsted jerseys, and 38 footballs. How many articles has he in all?

15. Eddie's brother saved up $15 to pay for a punching bag. It cost only $11.85. How much did he have left?

16. In the shoe store there are 58 shelves containing shoes. There are 18 pairs on each shelf. How many pairs in the store?

17. Helen's mother bought her a middy blouse for $1.75, a sport skirt for $5.50, and a pair of white shoes for $4.40. How much did she pay for the things?

18. Joe's brother works 8 hours a day, but last month he was paid for 192 hours of work. How many days' work was that?

19. If gasoline is 6 cents a quart, what will 2 gallons cost?

20. Mary's mother bought a suit at a sale for $28.75. Last week the same suit was marked $40. How much did she save by waiting for the sale?

21. From 8223 take 3679.

22. A lady handed a salesgirl a fifty-dollar bill to pay for an umbrella, which cost $3.45, and a pair of shoes, which cost $6.50. How much change should the lady receive?

ARITHMETIC BY GRADES

THIRD YEAR BOOK

SECOND HALF: GRADE 3B

I. READING AND WRITING NUMBERS

ORAL EXERCISE

Read:

1. 461	**2.** 9099	**3.** \$17.17	**4.** \$89.90
4061	2022	\$29.92	\$914.22
707	7077	\$601.77	\$1216.72
8982	1123	\$99.19	\$8800.00

Write in words:

5. \$27.16	**6.** \$13.75	**7.** 3404	**8.** 1118
\$4.50	\$506.69	679	8888
2022	3033	4044	6066

WRITTEN EXERCISE

Write in figures:

1. Three hundred five.
2. Three dollars and five cents.
3. Thirty three hundred eleven.

4. Thirteen hundred three.
5. Five thousand fifty.
6. Fifty-six dollars and five cents.
7. Three thousand five hundred fifty.
8. Thirty-three dollars and fifty-five cents.
9. Seventy-four dollars and seventy-four cents.
10. Six thousand dollars and forty-six cents.
11. Nine thousand seven hundred forty.
12. Ten thousand dollars.

ORAL EXERCISE: ADVANCED WORK

1. Count by thousands from 10,000 to 20,000, from 50,000 to 62,000; from 89,000 to 100,000.

2. Read:	20,000	50,000	70,000	80,000
	30,000	60,000	90,000	100,000
3. Read:	32,400	51,240	84,961	90,846
	46,700	66,372	89,307	99,273
	29,002	42,007	54,001	80,046
	37,009	10,009	60,084	73,004

WRITTEN EXERCISE: ADVANCED WORK

Write in figures:

1. Forty thousand.
2. Seventy thousand.
3. Thirty-three thousand seven hundred.
4. Fifty-nine thousand eight hundred six.
5. Seventy-four thousand eight hundred nine.
6. Eighty-six thousand seven hundred eleven.

7. Ninety-one thousand three hundred seventy-one.
8. One hundred thousand.
9. Twenty-five thousand four hundred forty-two.
10. Thirty-six thousand eight hundred ninety-eight.

Roman Numbers.

1. Read these numbers:

IV	IX	XVII	XX	XII	VII
VI	XI	XVI	VIII	XIV	X
V	XIX	XIII	III	XVIII	XV

2. Learn the following:

XXX=30 XL=40 L=50

3. Review by letters:

I=1 V=5 X=10 L=50

4. Review by tens:

X=10 XX=20 XXX=30 XL=40 L=50

ORAL EXERCISE

Read:

1.	XXIV	XXIX	XXIX	XLIV	XLIX
2.	XXV	XXX	XLI	XLVI	XLII
3.	XXVII	XXXVI	XXXIV	XLVIII	XLVII
4.	XXVI	XXXIII	XXXV	XLV	XLIII
5.	XXVIII	XXXI	XXXVIII	XL	L

WRITTEN EXERCISE

Write in Roman numbers:

1.	15	21	35	38	46	49
2.	19	24	33	40	41	42
3.	18	29	39	43	48	45

II. COUNTING

I. Count:

1. By 2's from 7 to 37.
2. By 3's from 4 to 40.
3. By 3's from 9 to 42.
4. By 4's from 2 to 50.
5. By 4's from 8 to 52.
6. By 4's from 4 to 56.
7. By 4's from 54 to 2.
8. By 5's from 66 to 11.
9. By 5's from 58 to 3.
10. By 6's from 67 to 7.
11. By 7's from 60 to 4.
12. By 6's from 6 to 66.

II. Count:

1. From 6 to 60 by 6's.
2. From 8 to 56 by 6's.
3. From 7 to 70 by 7's.
4. From 9 to 65 by 7's.
5. From 5 to 86 by 9's.
6. From 1 to 100 by 9's.
7. From 8 to 80 by 8's.
8. From 9 to 57 by 8's.
9. From 9 to 81 by 9's.
10. From 8 to 71 by 9's.
11. From 4 to 100 by 8's
12. From 6 to 83 by 7's.

III. 1. Beginning with 63, count by 6's to 3.
2. Beginning with 68, count by 7's to 5.
3. Beginning with 85, count by 7's to 8.
4. Beginning with 87, count by 6's to 9.

IV. Count:

1. From 103, by 5's to 3.
2. From 98, by 6's to 2.
3. From 88, by 7's to 4.
4. From 91, by 8's to 3.
5. From 94, by 7's to 3.
6. From 88, by 6's to 4.
7. From 82, by 9's to 1.
8. From 94, by 8's to 6.

III. ADDITION

ORAL EXERCISE

24	42	69	76	96	37	84	58

1. Add first 10 and then 4 to each of the above numbers.

2. Add first 8 and then 5 to each of the above numbers.

3. Add first 3 and then 9 to each of the above numbers.

Give sums rapidly:

4.	8	19	28	39	48	59	68	79
	3	3	3	3	3	3	3	3
5.	5	16	27	38	49	53	64	75
	6	6	6	6	6	6	6	6

Add:

6.	83	94	17	28	37	48	54	67
	6	6	6	6	6	6	6	6
7.	69	77	87	17	13	23	34	46
	6	6	6	6	6	6	6	6
8.	18	68	88	28	78	58	48	38
	9	9	9	9	9	9	9	9

Terms in Addition. The numbers to be added are called the **addends.**

The result in addition is called the **sum.**

We *check* the work by adding *first up and then down,* or *first down and then up.*

In adding dollars and cents, we write the decimal points underneath one another. We write the dollar sign before the sum.

WRITTEN EXERCISE

Add and check:

1.	**2.**	**3.**	**4.**
348	279	1563	$ 30.01
978	1844	472	536.25
79	516	1115	14.75
1484	2212	414	211.20

5.	**6.**	**7.**	**8.**
$54.62	975	546	373
6.95	666	666	666

9.	**10.**	**11.**	**12.**
322	$ 3.12	3470	2216
1849	14.17	665	737
735	9.50	519	456
300	516.21	2029	3003

13.	**14.**	**15.**	**16.**
611	547	3036	1117
2374	8296	755	549
596	138	432	3112
3411	68	1717	837
672	757	648	48

17.	18.	19.	20.
3247	$ 82.76	4645	2632
932	174.83	872	746
412	98.88	1575	511
3672	675.43	819	986

ORAL EXERCISE

Add 7 to each number: *Add 5 to each number:*

1.	12	22	42	92	62	52	72	32	82
2.	14	44	64	84	54	34	94	24	74
3.	25	65	75	45	35	55	95	85	15
4.	36	66	16	96	76	46	26	56	86

Add 4 to each number above: *Add 6 to each number above:*

Give sums:

5.	27	57	97	17	47	77	37	87	67
	7	7	7	7	7	7	7	7	7
6.	90	20	50	70	80	30	60	10	40
	40	40	40	40	40	40	40	40	40

Give answers rapidly:

7.	8.	9.	10.
88+8	18+8	84+8	72+8
77+8	19+8	73+8	83+8
66+8	41+8	15+8	53+8
55+8	52+8	29+8	16+8

11.	12.	13.	14.
44+8	63+8	38+8	27+8
33+8	74+8	78+8	39+8
22+8	85+8	47+8	49+8

WRITTEN EXERCISE

Add and test:

1.	2.	3.	4.
3494	$277.14	3069	6204
747	6.69	845	700
1492	532.43	79	1975
360	72.00	86	896
748	438.93	178	1170

5.	6.	7.	8.
345	$1000.04	3479	$222.38
1748	556.17	346	49.49
976	22.28	1400	117.68
843	4.69	738	98.04
1975	738.41	416	546.49

ORAL EXERCISE

Add 9 to each number:

1.	23	43	3	73	93	63	83	33	53
2.	18	78	8	38	88	68	48	28	58
3.	29	49	9	69	89	99	59	39	19

Give the sums rapidly:

4.	33	43	13	23	93	73	83	63	53
	8	8	8	8	8	8	8	8	8
5.	95	75	25	55	43	35	15	65	85
	8	8	8	8	8	8	8	8	8
6.	21	41	91	71	51	81	61	31	11
	8	8	8	8	8	8	8	8	8

Add:

7.	40	50	70	20	10	90	80	60	30
	80	80	80	70	70	70	60	60	60
8.	60	60	60	60	60	60	60	60	60
	35	15	45	25	65	55	70	85	95

9. Add 30 to each number: Add 50 to each number:

50	80	30	90	40	10	20	60	70

Add 90 to each number above: Add 40 to each number above.

Give sums rapidly:

10.	21	11	51	81	71	31	61	91
	9	9	9	9	9	9	9	9
11.	83	13	43	93	33	63	73	53
	9	9	9	9	9	9	9	9
12.	37	87	47	77	57	27	97	17
	9	9	9	9	9	9	9	9
13.	25	85	55	35	75	15	65	45
	9	9	9	9	9	9	9	9
14.	66	16	56	46	76	96	36	86
	9	9	9	9	9	9	9	9

WRITTEN EXERCISE

Add and check:

1.	2.	3.	4.
1307	$ 32.47	52	$ 5.00
846	3.16	549	15.18
201	500.08	3637	3.26
2777	22.00	748	111.49
369	19.97	1200	56.75

5.	6.	7.	8.
3467	4670	434	$ 26.14
208	685	6771	9.32
2375	3119	578	50.00
667	919	1209	4.18
1509	1288	548	226.68

9.	10.	11.	12.
5409	$ .69	6839	7138
617	4.18	438	193
1456	27.72	1473	567
739	338.45	234	408
3398	9.36	617	992

13.	14.	15.	16.
647	1400	389	$15.64
3049	398	1463	.82
968	643	587	7.16
564	4264	22	3.69
1082	181	439	4.25

17.	18.	19.	20.
$84.29	$141.13	$255.65	$775.12
13.43	75.	92.83	39.39
24.85	223.19	400.08	606.69

ORAL EXERCISE

20	60	40	90	70	10	50	80	100	30

1. Add 60 to each of the above numbers.
2. Add 20 to each number.
3. Add 10 to each number.
4. Add 70 to each number.
5. Add 30 to each number.
6. Add 40 to each number.
7. Add 50 to each number.

Give answers rapidly:

8.	9.	10.	11.	12.
19+9	38+8	16+7	15+6	17+4
29+9	59+8	36+7	75+6	37+4
49+9	89+8	56+7	85+6	77+4
69+9	69+8	76+7	88+6	7+4
39+9	19+8	86+7	25+6	47+4

WRITTEN EXERCISE

Add and check:

1.	2.	3.	4.
$122.04	4307	$ 14.33	1100
78.39	1654	6.49	8296
336.02	6611	5.11	2342
4.39	1772	127.96	5455
5.25	849	54.16	6768

5.	6.	7.	8.
5346	9327	$ 66.35	7300
8297	468	14.28	486
4004	793	164.72	1785
546	2067	561.03	1723
1311	2435	48.49	8243

Add and test results:

9.	10.	11.	12.
$ 42.25	1402	$ 62.48	4675
16.39	8463	7.39	8270
310.08	9876	14.15	5000
400.00	5672	73.25	4372
63.39	803	9.79	6088

13.	14.	15.	16.
$22.25	3405	6432	4385
8.35	3622	147	72
16.25	7826	82	1116
27.50	8374	937	238
14.25	5600	488	1072

17.	18.	19.	20.
$544.69	884	5485	$ 62.69
22.11	1273	987	3.49
5.48	987	336	7.54
16.67	546	47	11.23
3.25	63	505	6.25
114.75	211	49	104.25

21.	22.	23.	24.
1522	7742	6578	1199
5693	9709	4865	7705
5285	8256	6178	6662
4722	4818	6601	4735
8169	2129	7300	2969
5747	2593	8106	2633
4226	8426	5127	8565
9128	4120	33	7152

IV. SUBTRACTION

ORAL EXERCISE

Find the difference rapidly:

1.	70	30	60	20	40	50	80	90
	10	10	10	10	10	10	10	10
2.	93	13	23	33	43	53	63	83
	6	6	6	6	6	6	6	6

Read answers quickly:

3.	12	22	32	42	52	62	72	82
	−3	−6	−5	−3	−6	−5	−6	−6
4.	50	80	90	70	40	30	20	60
	−20	−20	−20	−20	−20	−20	−20	−20
5.	91	81	71	61	51	41	31	21
	−6	−6	−6	−6	−6	−6	−6	−6

Terms in subtraction.

43 *minuend*
27 *subtrahend*
16 *difference* or *remainder*

In subtraction the larger number is called the **minuend.** The smaller number is called the **subtrahend.** The result is called the **difference or remainder.**

Proof: 16 + 27 = 43.

ORAL EXERCISE

Supply the missing subtrahends:

1.	21	32	43	47	62	71	82	93
	?	?	?	?	?	?	?	?
	15	25	35	38	55	65	78	85
2.	90	80	70	60	40	20	30	50
	?	?	?	?	?	?	?	?
	70	60	50	40	20	20	10	30
3.	15	25	75	65	45	55	35	95
	?	?	?	?	?	?	?	?
	8	18	68	58	38	48	28	88

WRITTEN EXERCISE

Subtract and prove:

1. 438	2. $8.07	3. 5201	4. 474	5. 3635
89	5.78	776	365	377
6. 291	7. $8.61	8. 6002	9. 741	10. 7011
167	7.72	688	489	482
11. 328	12. $5.10	13. 4083	14. 460	15. 4616
177	4.71	389	367	378

16. From eighty-eight dollars and eleven cents, take nine dollars and seventy-seven cents.

17. From ten thousand four hundred six, take one thousand five hundred eighty-eight.

ORAL EXERCISE

Subtract:

1.	54	43	32	21	10	64	74	84	94
	8	8	8	8	8	8	8	8	8
2.	72	61	50	41	32	23	14	83	94
	9	9	9	9	9	9	9	9	9
3.	85	94	76	67	58	49	40	31	22
	7	7	7	7	7	7	7	7	7

Give the difference:

4. 50 − 20	**5.** 93 − 30	**6.** 70 − 50	**7.** 30 − 10
10 − 10	80 − 40	40 − 30	20 − 10
60 − 20	30 − 20	40 − 10	70 − 50

Supply the missing numbers:

8. 60 − ? = 10	**9.** 40 + ? = 80	**10.** ? − 50 = 80
50 − ? = 20	20 + ? = 40	20 + ? = 70
90 − ? = 70	30 + ? = 70	? − 50 = 90
10 + ? = 70	? − 30 = 70	50 + ? = 70
80 − ? = 50	30 + ? = 90	? − 40 = 60

State the difference:

11.	41	73	85	27	12	59	64	90	36
	8	8	8	8	8	8	8	8	8
12.	28	93	76	32	54	42	66	49	18
	8	8	8	8	8	8	8	8	8

Subtract:

13.	35	83	74	16	20	42	57	68	99
	8	8	8	8	8	8	8	8	8
14.	47	88	33	52	11	67	89	37	91
	8	8	8	8	8	8	8	8	8

WRITTEN EXERCISE

Subtract:

1. $522.65 484.78	**2.** 2823 743	**3.** $34.14 9.25	**4.** 4305 835
5. $500.65 69.74	**6.** 3716 2827	**7.** $34.12 25.38	**8.** 4106 997
9. $100.00 22.88	**10.** 1000 569	**11.** $56.24 19.38	**12.** 5467 3589
13. $262.25 48.76	**14.** 2002 844	**15.** 1646 978	**16.** 2269 1479
17. $72.16 38.39	**18.** 3017 288	**19.** 4302 3478	**20.** 2075 1886
21. $561.11 82.94	**22.** 7280 5934	**23.** 1119 329	**24.** 3200 884

What numbers added together will give the minuend? In each of the above examples, prove that your answer is correct.

ORAL EXERCISE

Give answers rapidly:

1.	40	70	90	60	30	10	20	50	80
	9	9	9	9	9	9	9	9	9
2.	22	92	72	32	52	62	42	82	12
	9	9	9	9	9	9	9	9	9

Subtract:

3.	35	85	75	15	25	45	55	65	95
	9	9	9	9	9	9	9	9	9
4.	47	27	37	57	17	67	77	87	97
	9	9	9	9	9	9	9	9	9

Subtract rapidly:

5.	16	36	56	76	96	46	26	66	86
	7	7	7	7	7	7	7	7	7
6.	12	22	92	72	52	32	42	62	82
	7	7	7	7	7	7	7	7	7

Tell the difference:

7.	33	93	43	13	53	63	83	73	23
	7	7	7	7	7	7	7	7	7
8.	28	18	58	98	88	38	48	78	68
	7	7	7	7	7	7	7	7	7

WRITTEN EXERCISE

1. The subtrahend is $711.38; the minuend is $598.49. Find the remainder and prove your answer.

2. The remainder is 1165; the subtrahend is 2276. Find the minuend.

Subtract and prove:

3. 2864 2475	4. $311.16 82.98	5. $22.10 8.25	6. 5247 1488
7. 7111 672	8. $111.69 35.75	9. 341 284	10. 4343 3434
11. 2100 995	12. $37.75 18.89	13. 634 585	14. 1360 496
15. ? 1834	16. $25.34 ?	17. ? 1172	18. 1669 ?

ORAL EXERCISE

Supply the missing minuends:

1.	?	?	?	?	?	?	?	?	?
	9	9	9	9	9	9	9	9	9
	41	77	54	28	86	65	32	71	99
2.	?	?	?	?	?	?	?	?	?
	6	6	6	6	6	6	6	6	6
	38	55	27	49	16	64	88	75	93

Supply the minuends rapidly:

3.	?	?	?	?	?	?	?	?	?
	7	7	7	7	7	7	7	7	7
	88	79	70	61	52	43	34	25	7

4.	?	?	?	?	?	?	?	?	?
	8	8	8	8	8	8	8	8	8
	14	36	74	56	94	84	66	24	46

Supply the missing numbers quickly:

5.	20	30	40	80	60	30	70	40	50
	−?	−?	−?	−?	−?	−?	−?	−?	−?
	10	20	20	50	20	10	32	10	30

6.	30	70	80	20	10	50	80	90	60
	−10	−40	−30	−10	−10	−40	−50	−50	−20

7.	?	?	?	?	?	?	?	?	?
	+8	+8	+8	+8	+8	+8	+8	+8	+8
	41	71	51	21	81	61	31	91	11

Give the difference:

8. 33−9	**9.** 13−9	**10.** 83−9	**11.** 63−9
93−9	43−9	23−9	73−9
12. 45−8	**13.** 65−8	**14.** 55−8	**15.** 45−8
16−7	76−7	26−7	46−7
66−7	96−7	56−7	86−7

Subtract rapidly:

16.	66	47	19	26	82	53	68	36	70
	9	9	9	9	9	9	9	9	9
17.	28	52	79	98	13	44	80	75	68
	8	8	8	8	8	7	7	7	7

Making Change. If you buy 69¢ worth of groceries and give the cashier a dollar bill, he says, "69, 70, 75, one dollar." As he says this, he hands you a cent, a nickel, and a quarter.

1. Count out the change for 25¢, when the purchase is:

14¢	22¢	19¢	11¢	7¢	16¢	13¢	3¢
6¢	17¢	16¢	9¢	12¢	8¢	18¢	21¢

2. Count out the change for 50¢, when the purchase is:

32¢	41¢	17¢	9¢	25¢	17¢	6¢	35¢
15¢	26¢	38¢	29¢	42¢	31¢	21¢	23¢

3. Count out the change for 75¢, when the purchase is:

55¢	62¢	66¢	53¢	71¢	52¢	57¢	61¢
51¢	69¢	67¢	58¢	74¢	69¢	59¢	54¢

4. Count out the change for $1, when the purchase is:

85¢	91¢	69¢	11¢	47¢	22¢	63¢	45¢
49¢	22¢	18¢	39¢	53¢	72¢	31¢	32¢

WRITTEN EXERCISE

Subtract:

1. 4265 3976	2. $30.14 28.65	3. 5005 966	4. $72.14 18.25
5. $3410 3399	6. 2498 999	7. 2000 1811	8. $364.00 229.92
9. $62.64 38.75	10. 1748 898	11. $111.11 39.49	12. 6103 5989
13. $172.12 84.69	14. 5251 4872	15. 6161 5979	16. 3412 3375
17. $89.20 29.48	18. $7615 877	19. $542.02 349.74	20. $161.15 59.78

What two numbers added together will equal the minuend?

In each of the above examples, prove that your answer is correct.

Subtract:

21. $82.07 8.28	22. 8206 3897	23. 9707 1829	24. $650.00 228.96
25. $7816.72 6409.83	26. 7800 6992	27. 8101 1756	28. $8179.85 936.69

ORAL DRILL EXERCISE

	A	B	C	D
	Tell the answers:			
1.	? 17 33	? 46 54	? 28 63	? 72 90
2.	80−30	40−10	90−70	70−30
3.	54 ? 28	84 ? 29	63 ? 37	71 ? 46
4.	5×7+1	6×7+3	7×7+4	6×4+1
5.	8×5+2	4×9+5	5×8+4	5×8+3
6.	54 +24 18 +37	34 +56 63 +73	67 +14 72 −19	43 −28 81 −12
	Multiply:			
7.	33 ×3 24 ×2	43 ×2 32 ×3	22 ×4 23 ×3	21 ×4 22 ×4
8.	23 ×4 12 ×3	33 ×2 14 ×3	44 ×2 13 ×3	32 ×4 13 ×3
	Divide:			
9.	21÷7	48÷12	49÷7	54÷6
10.	29÷7	36÷3	15÷3	42÷6

GENERAL ORAL PROBLEMS

1. We had 37 paint pans in our closet. Miss S. sent in 14 more. How many have we now?

2. We had 46 pupils in our class. 34 were promoted. How many were not promoted?

3. If it costs 15 cents each to join the athletic club, how much will it cost for 6 boys to join?

4. There are a dozen crayons in a box. How many boxes shall we need in order to give one crayon to each one of 60 pupils?

5. If our room is 27 feet wide, how many yards wide is it?

6. If a quart of oil costs 15 cents, what shall I have to pay for a gallon?

7. A peck of potatoes cost $.72. What will a quart of potatoes cost?

8. Sam left the house at 8 o'clock. He got to school three-quarters of an hour later. What time did he get to school?

9. The baby is 25 months old? How old is she in years and months?

10. $\frac{1}{2}$ pound of tea costs $.22. Find what 2 pounds of tea will cost.

11. I have 11 pounds of sugar. How many $\frac{1}{4}$ pound packages can I make up of it? How many $\frac{1}{2}$ pound packages? How many $\frac{1}{3}$ pound packages?

12. Milk costs 8 cents a quart. What will 12 quarts of milk cost?

GENERAL WRITTEN PROBLEMS

1. My father has $52.68 in the bank. If he takes out $49.79, how much will he leave in the bank?

2. The oilman bought $82.50 worth of oil. He has sold $69 worth. Find the value of the oil he has left.

3. The grocer has 32 gallons of oil. How many quart bottles can he fill?

4. Apples sell for $9.15 a barrel. What will 8 barrels cost?

5. A room has 6 rows of seats and 7 seats in each row. 8 seats are empty. How many pupils are present?

6. If each one of those present uses 3 pieces of paper, how many pieces of paper will they need in all?

7. We had $32.42 in our class treasury. Our glee club entertainment cost $14.59. How much have we left?

8. A man earns $55 a month. How much will he earn in a year?

9. In two classes containing 85 pupils, all but 36 lived more than two blocks away from the school. How many lived more than two blocks away?

10. Each window requires 16 panes of glass. How many panes will be needed for 12 windows?

11. We have 50 girls in the class. $\frac{1}{10}$ of them failed in spelling. How many passed in spelling?

12. Our spelling books cost 11 cents each. What will 45 cost?

V. MULTIPLICATION

Multiplying by 6.

1. $6+6+6=?$ How much is 3×6? $6\times3=?$
$6+6+6+6=?$ How much is 4×6? $6\times4=?$
$6+6+6+6+6=?$ How much is 5×6? $6\times5=?$

2. On your paper, complete the above table to 12×6.

The Table of 6's. Learn this:

$1\times6=6$	$7\times6=42$	$6\times1=6$	$6\times7=42$
$2\times6=12$	$8\times6=48$	$6\times2=12$	$6\times8=48$
$3\times6=18$	$9\times6=54$	$6\times3=18$	$6\times9=54$
$4\times6=24$	$10\times6=60$	$6\times4=24$	$6\times10=60$
$5\times6=30$	$11\times6=66$	$6\times5=30$	$6\times11=66$
$6\times6=36$	$12\times6=72$	$6\times6=36$	$6\times12=72$

ORAL EXERCISE

Multiply:

1.	2.	3.	4.	5.	6.
6×9	6×7	6×4	12×6	8×6	$6\times8+5$
6×3	6×12	6×10	4×6	9×6	$9\times6+4$
6×0	6×8	5×6	10×6	2×6	$7\times6+5$
6×11	6×5	11×6	3×6	7×6	$5\times6+6$
6×6	6×2	6×8	5×6	6×6	$4\times6+5$

7. Add 8 to 6×9.

8. Add 3 to 11×6.

9. Add 5 to 3×6.

10. Add 12 to 6×6.

11. Add 2 to 12×6.

12. Add 7 to 4×6.

8	12	4	3	11	9	10	2	6	7

13. Multiply each of the above numbers by 3; by 5.

14. Multiply each number by 6; by 4.

15. Multiply each number by 2; by 10.

16. What is the difference between the multiplier and the multiplicand?

Multiplying by 10.

70	20	30	80	60	50	40	10
×10	×10	×10	×10	×10	×10	×10	×10
700	200	300	800	600	500	400	100

To multiply these numbers by 10, we add a zero to the right of the number.

ORAL EXERCISE

Multiply:

1.	2.	3.	4.	5.
22×10	67×10	107×10	646×10	542×10
34×10	76×10	214×10	757×10	187×10
43×10	87×10	323×10	534×10	201×10
56×10	89×10	432×10	988×10	500×10

Multiplying by 20.

33	45	67	152	571	840	903
×20	×20	×20	×20	×20	×20	×20

To multiply by 20, 30, 40, etc., we bring down the zero to the right of the product and multiply the number by the tens' figure.

WRITTEN EXERCISE

Find the products:

1. 82 ×20	2. 96 ×30	3. 104 ×70	4. 18 ×30	5. 48 ×90
6. 39 ×40	7. 142 ×40	8. 58 ×40	9. 260 ×70	10. 89 ×20
11. 183 ×50	12. 24 ×20	13. 221 ×40	14. 135 ×60	15. 307 ×30
16. 500 ×40	17. 151 ×20	18. 89 ×30	19. 58 ×40	20. 93 ×50
21. 143 ×30	22. 220 ×40	23. 86 ×50	24. 79 ×80	25. 205 ×20

Multipliers of Two Figures.

If one suit of clothes costs $25, what will 23 suits cost?

$25 23 75 50 $575	We multiply 25 by 3 and write the product, 75, with the first figure, 5, under the 3. We multiply 25 by 2 and write the second product, 50, with the first figure, 0, under the 2. We add the two *partial products* and write the product $575.	PROOF $23 25 115 46 $575

WRITTEN EXERCISE

Multiply:

1. 22 21	2. 57 41	3. 25 13	4. 26 22	5. 183 40
6. 63 31	7. 63 33	8. 67 45	9. 144 50	10. 196 30
11. 24 51	12. 52 34	13. 58 35	14. 168 40	15. 121 42
16. 28 12	17. 34 42	18. 52 25	19. 175 50	20. 234 32
21. 37 16	22. 45 33	23. 24 75	24. 186 30	25. 332 24

Multiplying by 7.

1. $7+7=?$ $\quad 2\times7=?$ $\quad 7\times2=?$
   $7+7+7=?$ $\quad 3\times7=?$ $\quad 7\times3=?$
   $7+7+7+7=?$ $\quad 4\times7=?$ $\quad 7\times4=?$
2. On your paper, complete the above table to 12×7.

The Table of 7's. Learn this:

$1\times7=7$	$7\times7=49$	$7\times1=7$	$7\times7=49$
$2\times7=14$	$8\times7=56$	$7\times2=14$	$7\times8=56$
$3\times7=21$	$9\times7=63$	$7\times3=21$	$7\times9=63$
$4\times7=28$	$10\times7=70$	$7\times4=28$	$7\times10=70$
$5\times7=35$	$11\times7=77$	$7\times5=35$	$7\times11=77$
$6\times7=42$	$12\times7=84$	$7\times6=42$	$7\times12=84$

ORAL EXERCISE

4	6	8	7	3	9	10	11	12

1. Multiply each of the above numbers by 7 and add 5 to each product.

2. Multiply each number by 7 and add 6 to each product; add 3 to each product; add 4 to each product.

Give products rapidly:

3.	**4.**	**5.**	**6.**	**7.**
5×7	8×4	3×9	4×4	9×4
6×8	9×5	2×8	5×3	8×5
7×9	6×6	1×9	6×4	7×5
8×2	9×7	2×6	7×7	6×1
9×3	6×9	3×5	8×7	6×7

8. Add 2 to each product above.

9. Add 3 to each product above.

6	8	3	2	5	9	12	10	4	7	11

10. Multiply each of the above figures by 3 and add 2.

11. Multiply each figure by 5 and add 3.

12. Multiply each figure by 7 and add 4.

13. Multiply each figure by 6 and add 5.

14. Multiply each figure by 4 and add 3.

WRITTEN EXERCISE

Multiply:

1. 265	**2.** 874	**3.** 224	**4.** 847	**5.** 779
7	7	70	7	7
6. 577	**7.** 1205	**8.** 756	**9.** 568	**10.** 843
50	7	7	7	7
11. 917	**12.** 1405	**13.** 1432	**14.** 980	**15.** 471
7	7	7	7	40

16. 771×20 **17.** 377×20 **18.** 171×70

Find the products:

19. 227	**20.** 467	**21.** 172	**22.** 375	**23.** 217
30	60	12	16	21
24. 227	**25.** 86	**26.** 77	**27.** 57	**28.** 39
32	35	42	45	52
29. 88	**30.** 75	**31.** 97	**32.** 123	**33.** 49
55	63	65	68	72
34. 56	**35.** 128	**36.** 29	**37.** 34	**38.** 45
75	77	57	17	27
39. 58	**40.** 73	**41.** 84	**42.** 92	**43.** 86
37	47	57	67	38

44. 93×71 **45.** 79×32 **46.** 87×47

Multiplying by 8.

1. 8+8=?	2×8=?	8×2=?
8+8+8=?	3×8=?	8×3=?
8+8+8+8=?	4×8=?	8×4=?

2. On your paper, complete the above table to 12×8.

The Table of 8's. Learn this:

1×8= 8	7×8=56	8×1= 8	8× 7=56
2×8=16	8×8=64	8×2=16	8× 8=64
3×8=24	9×8=72	8×3=24	8× 9=72
4×8=32	10×8=80	8×4=32	8×10=80
5×8=40	11×8=88	8×5=40	8×11=88
6×8=48	12×8=96	8×6=48	8×12=96

ORAL EXERCISE

Give products rapidly:

1.	2.	3.	4.	5.	6.
3×8	2×8	10×8	8×10	8×12	8×3
9×8	4×8	12×8	8×5	8×2	8×6
7×8	11×8	1×8	8×4	8×7	8×8
5×8	6×8	8×8	8×11	8×1	8×9

Tell the results:

7.	8.	9.	10.	11.
3×8+3	5×8+4	6×8+5	7×8+6	8×8+7
9×8+7	4×8+6	2×8+5	1×8+4	5×8+3
8×2+4	8×5+3	8×8+2	8×9+3	8×7+6
8×1+7	8×3+5	8×4+4	8×6+2	8×8+3

10	5	7	4	9	12	8	11	3

12. Multiply each number above by 2; 7; 8; 4; 3.

13. Multiply each number by 6; add 3 to each product.

14. Multiply each number by 5; add 4 to each product.

15. Multiply each number by 8; add 7 to each product.

WRITTEN EXERCISE

Find products:

1. 1162 ×8	**2.** 1277 ×8	**3.** 865 ×8	**4.** 439 ×8	**5.** 1263 ×8
6. 283 ×80	**7.** 439 ×80	**8.** 654 ×80	**9.** 238 ×80	**10.** 139 ×80
11. 44 88	**12.** 54 88	**13.** 64 88	**14.** 74 88	**15.** 84 88
16. 188 76	**17.** 184 65	**18.** 182 44	**19.** 198 38	**20.** 128 28

21. Multiplier, 38; multiplicand, 245. Find the product.

22. Multiplier, 84; multiplicand, 166. Find the product.

23. Multiplier, 56; multiplicand, 252. Find the product.

24. Multiplier, 74; multiplicand, 189. Find the product.

25. 357×24 **26.** 638×28 **27.** 739×18 **28.** 546×43

Multiplying by 9.

1. Give the multiplication table of eights.

Give answers at sight:

2.	**3.**	**4.**	**5.**	**6.**
5×8+2	7×8+6	6×8+5	8×7+1	8× 6+5
3×8+3	4×8+7	8×3+6	8×9+2	8× 8+6
9×8+4	2×8+3	8×5+7	8×3+3	8×11+7
8×8+5	11×8+6	8×7+4	8×1+4	8×12+3

7. Count by 9 from 9 to 108.

8. 9+9=?	2×9=?	9×2=?
9+9+9=?	3×9=?	9×3=?
9+9+9+9=?	4×9=?	9×4=?

9. On your paper, complete the above table to 12×9.

The Table of 9's. Learn this:

1×9= 9	7×9=63	9×1= 9	9× 7=63
2×9=18	8×9=72	9×2=18	9× 8=72
3×9=27	9×9=81	9×3=27	9× 9=81
4×9=36	10×9=90	9×4=36	9×10=90
5×9=45	11×9=99	9×5=45	9×11=99
6×9=54	12×9=108	9×6=54	9×12=108

ORAL EXERCISE

Tell products at sight:

1.	2.	3.	4.	5.	6.
3×9	2×9	12×9	9× 9	9×6	9× 8
9×9	4×9	1×9	9× 3	9×5	9×11
7×9	11×9	8×9	9×10	9×1	9×12
6×9	5×9	9×2	9× 7	9×4	9× 0

Give answers rapidly:

7.	8.	9.	10.
4×9+7	6×9+6	11×9+2	2×9+1
5×9+5	8×9+8	10×9+5	3×9+3
7×9+4	12×9+3	1×9+8	9×9+8

8	3	6	2	7	9	11	4	10	12

11. Multiply each of the above numbers by 8; 5; 9; 6; 7.

12. Multiply each number by 4 and add 3.

13. Multiply each number by 2 and add 1.

14. Multiply each number by 3 and add 2.

WRITTEN EXERCISE

Multiply:

1.	2.	3.	4.	5.
345	748	956	58	114
×9	×9	×9	×90	×52

6.	7.	8.	9.	10.
132	67	156	119	96
29	90	75	34	88

Multiply:

11. 77	**12.** 108	**13.** 113	**14.** 234	**15.** 147
69	28	19	56	59
16. 83	**17.** 96	**18.** 112	**19.** 158	**20.** 158
79	99	90	44	39
21. 172	**22.** 99	**23.** 163	**24.** 467	**25.** 251
78	63	58	17	80

26. 216×32 **27.** 167×19 **28.** 155×26 **29.** 123×42

30. 334×27 **31.** 228×34 **32.** 108×39 **33.** 153×29

34. 254	**35.** 76	**36.** 291	**37.** 302	**38.** 128
23	49	62	28	39

39. There are 52 words on each page of the spelling book. If the book has 45 pages, how many words are there?

40. The rent of a store is $155 a month. Find the rent for a year.

41. If an errand boy works 9 hours a day, how many hours will he work in 31 days?

42. The grocer buys tomatoes in boxes that hold one dozen cans each. If he buys 84 boxes, how many cans will he have?

43. A storekeeper sells chairs for $12 each. How much will he receive for 97 chairs?

ORAL DRILL EXERCISE

	A	*B*	*C*	*D*
1.	$\frac{1}{6}$ of 54	$\frac{1}{5}$ of 80	$\frac{1}{4}$ of 88	$\frac{1}{7}$ of 77
2.	$\frac{1}{10}$ of 70	$\frac{1}{4}$ of 16	$\frac{1}{5}$ of 55	$\frac{1}{10}$ of 100
3.	$\frac{1}{2}$ of 28	$\frac{1}{8}$ of 64	$\frac{1}{6}$ of 66	$\frac{1}{5}$ of 75
4.	$\frac{1}{3}$ of 36	$\frac{1}{7}$ of 49	$\frac{1}{2}$ of 84	$\frac{1}{8}$ of 40
5.	$\frac{1}{5}$ of 35	$\frac{1}{6}$ of 36	$\frac{1}{10}$ of 90	$\frac{1}{3}$ of 24

Give answers:

6.	38+71	27+33	37−34	28+41
7.	79+31	84+18	93−18	29+12
8.	54+14	33+42	81−13	37+22
9.	63+15	56+17	48−14	46+24
10.	21+37	69+18	63−21	52+53

Multiply:

11.	8×8	8×6	7×7	9×9
12.	9×5	6×6	6×7	8×7
13.	6×3	3×8	7×9	8×4

Divide:

14.	28÷4	54÷6	42÷6	81÷9
15.	36÷4	24÷6	72÷8	27÷9

Add the following:

16.	$2.30	$4.20	$7.60	$8.50
	1.45	3.75	4.35	2.15
17.	$13.20	$14.80	$16.70	$11.10
	3.35	4.25	2.25	8.35

GENERAL ORAL PROBLEMS

1. How many 5-cent sandwiches can I buy for $.60?

2. Soldiers march 20 miles in one day. How far will they march in 4 days?

3. The class worked 54 examples. $\frac{1}{6}$ of them were wrong. How many were wrong?

4. George had 47 examples correct. I had 9 more correct than George. How many had I correct?

5. The grocer paid $28.60 for potatoes and $43.75 for oranges. How much more did the oranges cost than the potatoes?

6. He sells the oranges at 4¢ each. How many will he sell for 92 cents?

7. How many 5-cent stamps can I buy for $.75?

8. For bread and groceries Ethel paid $3.63. The bread bill was $.85. Find what the groceries cost.

9. How many 6-inch strips of ribbon can I cut from a piece 4 feet long?

10. Helen bought 8 yards of ribbon at $.12 a yard. How much change will she get from a 2-dollar bill?

11. We have 75 pads in the closet. How many will be left when we have used 55?

12. It is 2 o'clock by the school clock. How many minutes will pass before it is a quarter to three? 3 o'clock? 3.15?

13. Harold paid 3¢ for a pencil, 5¢ for a composition book, and 2¢ for an eraser. How much change will he receive from a quarter of a dollar?

GENERAL WRITTEN PROBLEMS

1. At $6 a ton, how many tons of coal can be bought for $96?

2. If we can seat 7 boys in each row of chairs, how many rows will be needed to seat 147 boys?

3. If a shirt costs $2, how many can I buy for $38?

4. How many 4-cent stamps will the clerk give out for 64¢?

5. John, the fruit dealer, pays $22.50 a month for rent. How much rent does he pay in a year?

6. In one month he bought $8.60 worth of cherries, and spent 7 times this amount for bananas. Find how much he paid for bananas.

7. He has saved $18.75 for a new stand. This will cost $46. How much more must he save?

8. A delivery wagon can carry 1425 pounds of groceries in one trip. How much can it carry in 8 trips?

9. A farmer had 648 quarts of milk. How many gallon cans could he fill from this?

10. In March, John sold $87.45 worth of fruit; in April, he sold $103.25 worth; and in May he sold $118.43 worth. How much did he sell in the three months?

11. Last week John paid $12.35 for bananas, $6.40 for apples, $3.80 for grapes, and $10.28 for peaches. How much did all the fruit cost him?

12. He sells oranges at 3 for 10¢. How much will he receive for 8 dozen at this rate?

VI. DIVISION

ORAL EXERCISE

1. If oranges are 4¢ each, how many can one buy for 32¢?

2. If there are 6 seats in a row, how many rows will 42 seats make?

3.	4.	5.	6.	7.
$18 \div 2$	$50 \div 2$	$20 \div 4$	$16 \div 8$	$21 \div 3$
$27 \div 3$	$30 \div 5$	$40 \div 5$	$33 \div 3$	$28 \div 4$
$55 \div 5$	$36 \div 3$	$60 \div 5$	$44 \div 11$	$45 \div 5$

WRITTEN EXERCISE

Divide:

1. 3)1230	**4.** 2)1422	**7.** 5)2055	**10.** 3)2439
2. 2)1824	**5.** 3)1563	**8.** 4)1600	**11.** 4)2048
3. 2)1042	**6.** 4)1244	**9.** 3)1839	**12.** 5)2550

Dividing by One Figure. Divide 9180 by 3:

```
3)9180
  3060
```

We write 3 directly under the 9 because there are 3 threes in 9.

We write 0 under 1 because there are no threes in 1.

We write 6 under 8 because there are 6 threes in 18.

We write 0 under 0 because there are no threes in 0.

Check: $3 \times 3060 = 9180$.

WRITTEN EXERCISE

Divide and check in the same way:

1. 212÷2	9. 3027÷3	17. 1624÷4
2. 816÷4	10. 1520÷5	18. 2515÷5
3. 912÷3	11. 1018÷2	19. 1214÷2
4. 540÷5	12. 1827÷3	20. 1632÷4
5. 1212÷4	13. 1218÷3	21. 6012÷2
6. 3535÷5	14. 3208÷4	22. 5015÷5
7. 2436÷4	15. 1644÷4	23. 1812÷3
8. 4530÷5	16. 2045÷5	24. 8244÷4

Dividing by 6.

1. I have 12 sheets of paper to give to 6 girls. How many sheets will each girl receive?

$12 \div 6 = ?$ $\frac{1}{6}$ of $12 = ?$ How many 6's in 12?

2. Blankbooks cost 6¢ each. How many can I get for 18¢?

$18 \div 6 = ?$ $\frac{1}{6}$ of $18 = ?$ How many 6's in 18?

3. There are 24 boys in the class. If we seat them 6 in a row, how many rows shall we have?

$24 \div 6 = ?$ $\frac{1}{6}$ of $24 = ?$ How many 6's in 24?

The Table of 6's. Learn this:

6÷6=1	42÷6= 7	$\frac{1}{6}$ of 6=1	$\frac{1}{6}$ of 42= 7
12÷6=2	48÷6= 8	$\frac{1}{6}$ of 12=2	$\frac{1}{6}$ of 48= 8
18÷6=3	54÷6= 9	$\frac{1}{6}$ of 18=3	$\frac{1}{6}$ of 54= 9
24÷6=4	60÷6=10	$\frac{1}{6}$ of 24=4	$\frac{1}{6}$ of 60=10
30÷6=5	66÷6=11	$\frac{1}{6}$ of 30=5	$\frac{1}{6}$ of 66=11
36÷6=6	72÷6=12	$\frac{1}{6}$ of 36=6	$\frac{1}{6}$ of 72=12

ORAL EXERCISE

1. Divide each number by 6:

6)24   0   12   60   42   18   30

2. Find $\frac{1}{6}$ of: 54; 6; 72; 66; 36; 48.

3.	?)42 7	6)? 8	7)? 5	12)? 4	7)42 ?
4.	?)54 9	7)? 6	8)? 5	6)? 9	8)48 ?
5.	?)44 11	?)48 6	?)54 6	6)36 ?	9)54 ?
6.	?)44 22	?)32 8	?)72 6	8)? 6	6)? 7

WRITTEN EXERCISE

Copy and divide:

1. 426÷6	9. 2530÷5	17. 1240÷4
2. 1212÷6	10. 1632÷4	18. 2733÷3
3. 1818÷6	11. 3642÷6	19. 1812÷2
4. 2412÷6	12. 3035÷5	20. 2436÷6
5. 1012÷2	13. 3648÷6	21. 3648÷6
6. 912÷3	14. 3545÷5	22. 4440÷4
7. 1216÷4	15. 6624÷6	23. 2430÷6
8. 2025÷5	16. 4424÷4	24. 3045÷5

Find the quotients:

25. 2412÷6	29. 4428÷4	33. 2050÷5
26. 3208÷8	30. 3545÷5	34. 5454÷9
27. 2128÷7	31. 3606÷6	35. 3248÷8
28. 4554÷9	32. 4235÷7	36. 2142÷7

Division with a Remainder.

I have a piece of ribbon 9 inches wide. How many pieces 2 inches wide can I cut from it?

2)9 4 and 1 left over.	Here we find that 2 is contained in 9 four times and that 1 is left over. This number left over is called a remainder.

ORAL EXERCISE

1. Tell how much is left over if we divide these numbers by 5:

6	11	17	21	26	28	31	36
9	13	19	24	33	37	41	52

2. Tell how much is left over if we divide these numbers by 4:

8	7	6	5	9	11	17	21
15	29	13	18	30	33	45	41

Give at sight the remainders in the following examples:

3.	2)5	2)9	2)7	2)8	3)7	3)13	3)8
4.	3)23	4)5	4)9	4)14	5)11	5)17	6)13
5.	3)14	3)17	6)26	6)27	6)21	6)25	6)14
6.	3)32	4)15	6)45	2)25	5)24	3)19	2)21
7.	5)46	3)29	5)58	4)19	3)25	6)37	4)49

WRITTEN EXERCISE

1. I have 546 inches of wire. How many 3-inch pieces can I cut from it?

$3\overline{)546}$ 182	There is 1 three in 5, and 2 left over. We write the 1 under the 5. The 2 hundreds added to the 4 tens give 24 tens. There are 8 threes in 24. We write the 8 under the 4. There are 2 threes in 6. We write the 2 under the 6.

Copy and divide in the same way:

2. $2\overline{)324}$	**6.** $3\overline{)557}$	**10.** $4\overline{)564}$	**14.** $5\overline{)655}$
3. $2\overline{)544}$	**7.** $3\overline{)549}$	**11.** $4\overline{)528}$	**15.** $5\overline{)600}$
4. $2\overline{)598}$	**8.** $3\overline{)516}$	**12.** $4\overline{)964}$	**16.** $5\overline{)705}$
5. $2\overline{)7212}$	**9.** $3\overline{)726}$	**13.** $4\overline{)960}$	**17.** $5\overline{)755}$

ORAL EXERCISE

1. Divide each number by 5 and give the remainders:

12 17 21 26 27 32 44 37

2. Divide these by 6 and give remainders:

26 37 50 44 23 17 25 33

3. Divide these by 4 and give the remainders:

22 29 14 17 25 22 18 11

4. Give the missing numbers:

$42 \div ? = 6$	$? \div 7 = 5$	$56 \div ? = 6$	$36 \div ? = 6$	$? \div 11 = 6$
$? \div 9 = 6$	$24 \div ? = 12$	$42 \div ? = 7$	$40 \div ? = 8$	$? \div 3 = 12$

Dividing by 7.

1. I have 21 one-cent stamps. How many letters can I mail with them if each letter takes 7 stamps?

$7\overline{)21}$? $21 \div 7 = ?$ $\frac{1}{7}$ of $21 = ?$ How many 7's in 21?

2. Each pupil needs 7 crayons. I have 35 pieces. Find how many pupils I can supply.

$7\overline{)35}$? $35 \div 7 = ?$ $\frac{1}{7}$ of $35 = ?$ How many 7's in 35?

The Table of 7's. Learn this:

$7 \div 7 = 1$	$49 \div 7 = 7$	$\frac{1}{7}$ of $7 = 1$	$\frac{1}{7}$ of $49 = 7$
$14 \div 7 = 2$	$56 \div 7 = 8$	$\frac{1}{7}$ of $14 = 2$	$\frac{1}{7}$ of $56 = 8$
$21 \div 7 = 3$	$63 \div 7 = 9$	$\frac{1}{7}$ of $21 = 3$	$\frac{1}{7}$ of $63 = 9$
$28 \div 7 = 4$	$70 \div 7 = 10$	$\frac{1}{7}$ of $28 = 4$	$\frac{1}{7}$ of $70 = 10$
$35 \div 7 = 5$	$77 \div 7 = 11$	$\frac{1}{7}$ of $35 = 5$	$\frac{1}{7}$ of $77 = 11$
$42 \div 7 = 6$	$84 \div 7 = 12$	$\frac{1}{7}$ of $42 = 6$	$\frac{1}{7}$ of $84 = 12$

ORAL EXERCISE

1. Find $\frac{1}{7}$ of:

35	28	7	49	77	56
21	14	84	42	63	70

State the results:

2.	3.	4.	5.	6.	7.
7×6	4×7	7×9	5×7	7×3	7×7
9×7	8×7	2×7	3×7	6×7	7×4
$77 \div 7$	$72 \div 7$	$28 \div 7$	$14 \div 7$	$35 \div 7$	$21 \div 7$
$56 \div 7$	$63 \div 7$	$70 \div 7$	$42 \div 7$	$49 \div 7$	$7 \div 7$
7×5	7×8	$84 \div 7$	7×10	12×7	7×11

Supply the missing multiplicands:

8.	?	?	?	?	?	?	?	?	?
	7	7	7	7	7	7	7	7	7
	63	77	7	35	42	14	84	21	28
9.	?	?	?	?	?	?	?	?	?
	6	6	6	6	6	6	6	6	6
	42	60	18	48	72	66	24	36	30

Supply the multipliers:

10.	7	7	7	7	7	7	7	7	7
	?	?	?	?	?	?	?	?	?
	14	49	70	84	63	21	35	28	77
11.	3	3	3	3	3	3	3	3	3
	?	?	?	?	?	?	?	?	?
	24	12	9	15	36	27	30	18	6

ORAL EXERCISE

1. Divide each number by 7 and give remainder:

17 58 65 36 59 13 47 52 38

67 45 23 9 48 18 39 33 19

2. Divide by 4 and give the remainder:

38 13 22 35 30 15 7 27 33

3. Divide by 6 and give the remainder:

15 39 57 49 33 14 26 39 8

4. Divide by 5 and give the remainder:

22 49 34 18 9 14 29 41 17

WRITTEN EXERCISE

Divide:

1. 5646÷3	5. 7644÷6	9. 7895÷5	13. 8088÷4
2. 2821÷7	6. 3549÷7	10. 8435÷7	14. 1414÷7
3. 2135÷7	7. 4207÷7	11. 5628÷7	15. 4921÷7
4. 3125÷7	8. 2471÷7	12. 1694÷7	16. 1854÷7

ORAL EXERCISE

Tell how much is left over:

1.	2.	3.	4.	5.	6.	7.
7)45	6)28	5)39	4)21	3)29	7)67	5)42
7)60	6)33	5)29	4)27	7)39	6)59	5)29

Dividing by 8.

1. I have 32¢ with which to buy stamps. How many letters can I mail if each letter takes 8¢ postage?

8)32   32÷8=? $\frac{1}{8}$ of 32=?   How many 8's in 32?
?

2. Each girl needs 8 sheets of paper. I have 48 pieces. How many girls can I supply?

8)48   48÷8=? $\frac{1}{8}$ of 48=?   How many 8's in 48?

The Table of 8's. Learn this:

8÷8=1	56÷8= 7	$\frac{1}{8}$ of 8=1	$\frac{1}{8}$ of 56= 7
16÷8=2	64÷8= 8	$\frac{1}{8}$ of 16=2	$\frac{1}{8}$ of 64= 8
24÷8=3	72÷8= 9	$\frac{1}{8}$ of 24=3	$\frac{1}{8}$ of 72= 9
32÷8=4	80÷8=10	$\frac{1}{8}$ of 32=4	$\frac{1}{8}$ of 80=10
40÷8=5	88÷8=11	$\frac{1}{8}$ of 40=5	$\frac{1}{8}$ of 88=11
48÷8=6	96÷8=12	$\frac{1}{8}$ of 48=6	$\frac{1}{8}$ of 96=12

ORAL EXERCISE

Give answers rapidly:

1.	2.	3.	4.	5.
88÷8	$\frac{1}{8}$ of 72	$\frac{1}{8}$ of 80	64÷8	$\frac{1}{8}$ of 96
8÷8	56÷8	40÷8	$\frac{1}{8}$ of 88	80÷8
16÷8	$\frac{1}{8}$ of 56	$\frac{1}{8}$ of 16	96÷8	$\frac{1}{8}$ of 32
24÷8	48÷8	72÷8	32÷8	$\frac{1}{8}$ of 24

WRITTEN EXERCISE

Divide:

1. $7\overline{)2254}$	6. $6\overline{)8506}$	11. $7\overline{)3003}$	16. $6\overline{)7104}$
2. $7\overline{)329}$	7. $8\overline{)208}$	12. $8\overline{)4016}$	17. $8\overline{)4248}$
3. 8808÷8	8. 6456÷8	13. 4872÷8	18. 4928÷8
4. 7000÷8	9. 8212÷4	14. 1176÷7	19. 8816÷8
5. 9992÷8	10. 2955÷3	15. 4818÷6	20. 5232÷8

Finding Factors of Numbers. We call 3 and 7 the **factors** of 21 because when *multiplied together*, they make 21.

1. Give the numbers which when multiplied together will equal the following:

14	36	81	24	108	54
25	42	48	60	72	55

2. When 3 is one factor, give the other factor of 9; 15; 21; 33; 30; 27; 3; 12.

3. Give two factors of each of the following numbers:

27	48	55	54	63	80	45	22

Dividing by 9.

1. I have 27 sheets of drawing paper to distribute among the pupils. How many girls can I supply with paper if I give each one 9 sheets?

$9\overline{)27}$? $\qquad 27 \div 9 = ? \qquad \frac{1}{9}$ of $27 = ?$

2. My father's packages each need 9 cents in stamps. How many can I mail if I have 36 cents to buy postage?

$9\overline{)36}$? $\qquad 36 \div 9 = ? \qquad \frac{1}{9}$ of $36 = ?$

The Table of 9's. Learn this:

$9 \div 9 = 1$	$63 \div 9 = 7$	$\frac{1}{9}$ of $9 = 1$	$\frac{1}{9}$ of $63 = 7$
$18 \div 9 = 2$	$72 \div 9 = 8$	$\frac{1}{9}$ of $18 = 2$	$\frac{1}{9}$ of $72 = 8$
$27 \div 9 = 3$	$81 \div 9 = 9$	$\frac{1}{9}$ of $27 = 3$	$\frac{1}{9}$ of $81 = 9$
$36 \div 9 = 4$	$90 \div 9 = 10$	$\frac{1}{9}$ of $36 = 4$	$\frac{1}{9}$ of $90 = 10$
$45 \div 9 = 5$	$99 \div 9 = 11$	$\frac{1}{9}$ of $45 = 5$	$\frac{1}{9}$ of $99 = 11$
$54 \div 9 = 6$	$108 \div 9 = 12$	$\frac{1}{9}$ of $54 = 6$	$\frac{1}{9}$ of $108 = 12$

ORAL EXERCISE

1.	2.	3.	4.	5.
$90 \div 9$	$\frac{1}{9}$ of 81	$36 \div 9$	$\frac{1}{9}$ of 9	$\frac{1}{9}$ of 27
$54 \div 9$	$63 \div 9$	$45 \div 9$	$108 \div 9$	$81 \div 9$
$72 \div 9$	$\frac{1}{9}$ of 18	$\frac{1}{9}$ of 108	$99 \div 9$	$\frac{1}{9}$ of 45

Divide each of these numbers by 9 *and give the remainder:*

6.	7.	8.	9.	10.
68	10	12	60	65
77	20	50	43	75

WRITTEN EXERCISE

1. 5)6010
2. 6)7212
3. 7)8428
4. 9)4518
5. 9)7236
6. 8)1608
7. 5)5560
8. 7)8414
9. 8)9608
10. 3)5403
11. 3)7296
12. 5)1520
13. 8)6728
14. 8)5840
15. 9)7650
16. 9)6489
17. 9)8469
18. 7)7007
19. 9)9027
20. 8)9616
21. 8)4968
22. 8)9672
23. 7)2107
24. 6)1812

Find the quotient when:

25. The dividend is 8435 and the divisor 7.
26. The dividend is 6568 and the divisor 8.
27. The dividend is 2439 and the divisor 9.
28. The dividend is 3780 and the divisor 9.
29. The dividend is 1830 and the divisor 6.
30. The dividend is 4260 and the divisor 6.

Divide:

31. 2592÷3
32. 3787÷9
33. 4156÷4
34. 5685÷5
35. 5407÷7
36. 3105÷9
37. 6198÷6
38. 9192÷8
39. 5894÷7
40. 6944÷7
41. 6183÷9
42. 7374÷3
43. 8296÷4
44. 9474÷6
45. 8604÷9
46. 4492÷7
47. 8928÷9
48. 7176÷8
49. 8532÷3
50. 4293÷9
51. 1197÷9
52. 2268÷9
53. 3384÷9
54. 4196÷4
55. 9117÷9
56. 5598÷9
57. 1098÷9
58. 2185÷5
59. 4396÷7
60. 7596÷9

ORAL DRILL EXERCISE

	A		B		C		D	
1.	63 −17	86 −24	72 −36	95 −28	41 −32	84 −56	54 −48	32 −18

	A	B	C	D
2.	70−50	60−10	80−60	60−30
3.	90−30	80−40	90−50	70−60

Tell the answers:

	A	B	C	D
4.	?)77 11	7)? 4	6)? 7	?)45 9
5.	?×11=66	54÷?=9	6×7	?÷3=6
6.	96÷?=12	8×3	4×11	49÷?=7
7.	$\frac{1}{8}$ of 64	$\frac{1}{7}$ of 42	$\frac{1}{2}$ of 44	$\frac{1}{9}$ of 81
8.	$\frac{1}{4}$ of 48	$\frac{1}{8}$ of 48	$\frac{1}{6}$ of 36	$\frac{1}{5}$ of 55
9.	$\frac{1}{3}$ of 45	$\frac{1}{5}$ of 60	$\frac{1}{9}$ of 72	$\frac{1}{7}$ of 28

Give quotients:

	A	B	C	D
10.	4)128	6)126	4)164	8)168
11.	5)105	7)147	7)287	5)250
12.	6)186	6)246	3)369	9)99

Subtract the following:

	A	B	C	D
13.	\$9.54 .28	\$5.45 .56	\$6.73 .80	\$8.45 .95
14.	\$11.63 .49	\$8.45 .56	\$9.20 .32	\$10.18 .80

GENERAL ORAL PROBLEMS

1. Tea is bought for $.40 a pound and sold for $.32 a pound. How much money is lost on each pound?

2. Cocoa is sold for 19¢ a box. The grocer makes 6¢ on each box. How much does he pay for it?

3. Our rent now is $36 a month. Before we moved, we paid $25. How much more do we pay now?

4. The butcher gained 8¢ a pound on his meat. He sold it for $.37 per pound. How much did he pay a pound?

5. Our phonograph cost $25. We sold it at a loss of $5.50. How much did we get for it?

6. Oranges sell at 3 for 10¢. What will 1 dozen cost? How many times 3 is 1 dozen?

7. Potatoes are selling at 3 quarts for a quarter. What will 9 quarts cost? How many times 3 is 9?

8. Rice is sold at 2 pounds for 15 cents. What will 8 pounds cost?

9. I arranged 42 boys in rows of 7. How many rows of boys did I have?

10. Van mailed 14 letters yesterday and 11 to-day. How many did he mail?

11. $\frac{1}{5}$ of his letters were sent to Brooklyn. How many were sent to Brooklyn?

12. I bought 5 gallons of oil. How many quarts did I buy? How many pints?

13. The oil cost $.04 a pint. How many quarts could I get for $.64?

GENERAL WRITTEN PROBLEMS

1. If we spend 5 hours a day in school, how many hours do we spend there in 23 days?

2. The fruit man bought 18 doz. bananas at $.09 a doz. How much did he pay for them?

3. Last week all his fruit cost him $26.40. He sold it at a profit of $13.45. For how much did he sell it?

4. The grocer bought apples at $7 a barrel. He spent $98. How many barrels did he get? Last week he sold 56 quarts of potatoes. How many pecks was that?

5. If a canal boat can carry 897 tons of coal a trip, how much coal can it carry in 8 trips?

6. I have 126 pictures to tie up in packages of 6 pictures each. How many packages shall I make?

7. Rhubarb sells at 3¢ a bunch. How many bunches can the restaurant man buy for 87 cents?

8. Last week he bought it at 4 bunches for 10¢. How much did he pay for 16 bunches?

9. He uses 5 dozen eggs every day. How many days will 125 dozen last him?

10. How many 3¢ oranges can he buy for 72 cents?

11. He uses 4 napkins on each table. With 128 napkins, how many tables can he supply?

12. He has 60 tables. $\frac{1}{3}$ of them are square. How many tables are square?

13. He hires 28 waiters. $\frac{1}{4}$ of them are not working. How many are not working?

Long Division. Divisors of Two Figures.

Divide 2748 by 12:

$$\begin{array}{r} 229 \\ 12\overline{)2748} \\ \underline{24} \\ 34 \\ \underline{24} \\ 108 \\ \underline{108} \end{array}$$

(1) Since 12 is not contained in 2, we *divide* 27 by 12. We write 2 in the quotient above the 27.

(2) Then we *multiply* 12 by 2. We *subtract* 24 from 27, and bring down the next figure, 4. 34÷12=2. Again we multiply, subtract, and bring down the next figure, 8. 108÷12=9.

Proof: 229×12=2748.

In the example above, 2748 is the **dividend,** 12 the **divisor,** and 229 the **quotient.**

WRITTEN EXERCISE

Divide in the same way:

1. 462÷21	**13.** 288÷24	**25.** 737÷11
2. 182÷14	**14.** 224÷16	**26.** 375÷25
3. 382÷28	**15.** 621÷27	**27.** 3960÷30
4. 416÷32	**16.** 442÷34	**28.** 589÷31
5. 864÷72	**17.** 990÷30	**29.** 891÷81
6. 748÷22	**18.** 588÷21	**30.** 781÷71
7. 7828÷38	**19.** 4925÷26	**31.** 2695÷39
8. 5568÷32	**20.** 6075÷81	**32.** 4920÷41
9. 9856÷28	**21.** 2700÷87	**33.** 2420÷59
10. 5624÷38	**22.** 3910÷85	**34.** 1269÷42
11. 2944÷32	**23.** 4615÷71	**35.** 2775÷39
12. 2304÷36	**24.** 3455÷75	**36.** 6606÷31

VII. FRACTIONS

Finding Two Thirds.

1. If you wish to find $\frac{1}{3}$ of anything, by what number do you divide?

2. If you have 12 pencils and you give $\frac{1}{3}$ of them to the class, how many do you give out?

3. If you wished to give two thirds of them to the class, how many would you give out?

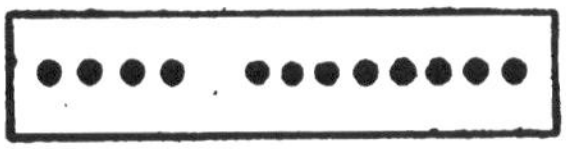

4. How many times $\frac{1}{3}$ is $\frac{2}{3}$?

5. What is $\frac{1}{3}$ of 12? Of 9? Of 15? Of 21?

6. What is $\frac{2}{3}$ of 12? Of 9? Of 15? Of 21?

Notice that $\frac{2}{3}$ of a number is $\frac{1}{3}$ of it multiplied by 2.

ORAL EXERCISE

1. Find $\frac{1}{3}$ of these numbers:

18 15 24 27 30 33 6 9 36

2. Find $\frac{2}{3}$ of the numbers above.

3. Find $\frac{1}{2}$ of: 16; 18; 30; 44; 50; 60; 80; 20.

4. Find $\frac{1}{3}$ of the following, and then $\frac{2}{3}$ of each:

90 feet	30 days	27 dollars	33 cents
36 inches	21 years	12 yards	60 minutes

5. Find $\frac{1}{4}$ of: 36; 12; 24; 8; 40; 28; 44; 4; 16.

6. Find $\frac{1}{5}$ of: 5; 45; $.35; 60; 55; 10; $25; 40; 50.

7. Find $\frac{1}{6}$ of: 12; 24; 36; 60; 66; 42; 48; 6; 30.

8. Find $\frac{1}{8}$ of: 8; 16; 32; 40; 24; 48; 56; 72; 64.

ORAL EXERCISE

1. How many inches in $\frac{1}{2}$ of a foot? $\frac{1}{3}$ of a foot? $\frac{1}{4}$ of a foot? $\frac{1}{6}$ of a foot?

2. How many ounces in $\frac{1}{2}$ of a pound? $\frac{1}{8}$ of a pound? $\frac{1}{4}$ of a pound?

3. How many months in $\frac{1}{2}$ of a year? $\frac{1}{4}$ of a year? $\frac{1}{3}$ of a year? $\frac{1}{6}$ of a year?

4. How many quarts in $\frac{1}{2}$ of a gallon? $\frac{1}{4}$ of a gallon?

5. How many minutes in $\frac{1}{2}$ of an hour? $\frac{1}{3}$ of an hour? $\frac{1}{5}$ of an hour? $\frac{1}{4}$ of an hour? $\frac{1}{6}$ of an hour? How many hours in $\frac{1}{6}$ of a day?

WRITTEN EXERCISE

1. If the grocer had 840 cans of corn, and sold $\frac{1}{2}$ of them, how many cans did he sell?

2. A man had $448. He put $\frac{1}{4}$ of it in the bank. How much did he put in the bank?

3. We have 330 pupils on our floor. $\frac{2}{3}$ of them are in the assembly room. How many are in the assembly room?

4. There were 854 books in the basement. The janitor sent up $\frac{1}{7}$ of them. How many are upstairs now?

5. My father had $368. He bought a cow with $\frac{1}{8}$ of this sum. How much did the cow cost?

6. How much would 12 cows cost at the same price?

Finding Three Quarters.

1. Find $\frac{1}{4}$ of one foot on your ruler. How many inches is this?

2. Count three of these quarters. How many inches?

3. $\frac{1}{4}$ of 12? $\frac{3}{4}$ of 12?

4. When $\frac{1}{4}$ is given, how do you find $\frac{3}{4}$?

5. Find $\frac{3}{4}$ of 12; 40; 20; 28; 16; 4; 24; 32; 8.

6. Find $\frac{3}{4}$ of:

$24	40 apples	16 rulers	48 stamps
20 pencils	32 quarts	8 berries	28 cards
16 pictures	36 words	44 examples	4 dollars

7. There are 36 pupils in the yard, and $\frac{3}{4}$ of them are boys. How many boys are there?

8. How many minutes are there in $\frac{1}{4}$ of an hour? In $\frac{3}{4}$ of an hour?

ORAL EXERCISE

1. Find $\frac{1}{3}$ of: 27; 36; 15; 3; 9; 30; 90; 24; 18.
2. Find $\frac{1}{2}$ of: 36; 48; 66; 32; 24; 50; 88; 100; 60.
3. Find $\frac{2}{3}$ of: 18; 24; 33; 45; 30; 27; 12; 9; 3; 90.
4. Find $\frac{1}{4}$ of: 8; 28; 40; 80; 100; 48; 44; 16; 60.
5. Find $\frac{3}{4}$ of: 20; 24; 28; 40; 16; 48; 36; 100; 60.
6. Find $\frac{1}{5}$ of: 50; 5; 100; 35; 55; 60; 20; 15; 45.
7. Find $\frac{1}{6}$ of: 12; 66; 60; 24; 6; 30; 18; 48; 36; 42.
8. Find $\frac{1}{8}$ of: 40; 56; 72; 96; 88; 48; 16; 64; 24; 32.
9. Find $\frac{1}{10}$ of: 20; 40; 90; 70; 50; 30; 80; 60; 100.

ORAL DRILL EXERCISE

	A	*B*	*C*	*D*
	Tell the answers:			
1.	49+11	28÷7	96−45	$\frac{1}{8}$ of 48
2.	54÷6	4×12	12×5	93−17
3.	9×8	45−17	77÷11	$\frac{1}{3}$ of 27
4.	66−18	$\frac{1}{7}$ of 42	32−14	6×7
5.	$\frac{1}{8}$ of 64	63÷7	$\frac{1}{5}$ of 100	48÷4
	Divide:			
6.	5)12	5)17	7)22	5)24
7.	84÷12	66÷11	54÷9	48÷4
8.	5)26	8)33	9)20	4)17
9.	36÷4	55÷5	81÷9	45÷5
10.	3)20	5)31	9)30	8)25
	Give answers:			
11.	72+18	$\frac{1}{5}$ of 60	53−27	$\frac{1}{2}$ of 200
12.	60÷12	96+14	7×20	6×30
13.	8×40	50÷5	$\frac{1}{3}$ of 66	76+28
14.	76−32	$\frac{1}{4}$ of 100	90÷3	62−14
15.	$\frac{1}{6}$ of 48	38−17	$\frac{1}{6}$ of 72	100÷2
	Answer the following:			
16.	6 doz. = ?	3 yr. = ? mo.	$\frac{1}{4}$ yd. = ? in.	3 bu. = ? pk.
17.	2 yd. = ? ft.	5 wk. = ? da.	3 min. = ? sec.	$\frac{1}{2}$ lb. = ?oz.
18.	2 gal. = ? pt.	2 da. = ? hr.	12 pt. = ? qt.	$\frac{1}{2}$ doz. = ?
19.	3 hr. = ? min.	4 doz. = ?	32 oz. = ? lb.	$1\frac{1}{2}$ ft. = ? in.
20.	3 pk. = ? qt.	$\frac{1}{2}$ yd. = ? in.	3 gal. = ? qt.	1 bu. = ? qt.

GENERAL ORAL PROBLEMS

1. I paid $.08 for a pencil box. I sold it for $.15. How much did I make by this sale?

2. There are 8 boys in each of 6 rows in our room. How many boys in the room?

3. $\frac{1}{6}$ of our class of 42 girls were promoted. How many were promoted?

4. I lost $.50 when I sold my skates, which cost $2. How much did I get for them?

5. Oil sells for $.20 a quart. How much will 4 gallons cost?

6. We have 40 paint brushes in our closet which is 6 more than the number of paint pans. How many paint pans have we?

7. At $.40 a dozen, how many dozen oranges can be bought for $2?

8. Helen is 9 years old and her brother is 8 years older. How old is her brother?

9. John has 15 examples right. How many more must he work correctly to have 27 right?

10. If 3 oranges cost $.12, what will 9 cost?

11. At the same rate, what will 24 oranges cost? How many times 3 is 24?

12. We use 5 eggs a day in our house. In how many days will we use 75 eggs?

13. Our class had 32 tickets for Field Day. $\frac{1}{4}$ were morning tickets. How many were morning tickets?

VIII. MEASURES

Dry Measure.

1. Suppose you are measuring peas. How many times will you have to fill the pint measure to make one quart?

2. How many times must you fill the quart measure to make one peck?

3. How many times do you fill the peck measure to make a bushel?

4. Name five other vegetables the grocer sells by this measure.

Table of Dry Measure.

2 pints (pt.)	=1 quart (qt.)
8 quarts	=1 peck (pk.)
4 pecks	=1 bushel (bu.)

ORAL EXERCISE

1. 2 pints =? $\frac{1}{2}$ of a quart=? A pint is what part of a quart?

2. 8 qt.=? $\frac{1}{8}$ pk.=? A qt. is what part of a pk.?

3. 4 pecks=? $\frac{1}{4}$ of a bushel=? A peck is what part of a bushel?

4. How many times would you fill the pint measure to make 16 quarts?

5. How many times would you fill the quart measure to make 2 pecks?

6. How many times would you fill the peck measure to make 2 bushels?

WRITTEN EXERCISE

1. The grocer has 2 pecks of apples. How many quart baskets can he fill?

Change:

2. 3 pk. to qt.
3. $\frac{1}{2}$ pk. to qt.
4. 9 bu. to pk.
5. 5 bu. to pk.
6. 7 qt. to pt.
7. 19 qt. to pt.

8. In 18 pints, how many quarts of walnuts has the grocer?

Change:

9. 9 pt. to qt.
10. 22 pt. to qt.
11. 11 qt. to pk.
12. 19 qt. to pk.
13. 8 pk. to bu.
14. 25 pk. to bu.

Liquid Measure.

1. How many times do you have to fill the pint measure with water to make one quart? To make 2 quarts? To make one gallon?

2. How many times do you have to fill the quart measure to make one gallon? To make 3 gallons?

Table of Liquid Measure.

2 pints (pt.)	= 1 quart (qt.)
4 quarts	= 1 gallon (gal.)

ORAL EXERCISE

1. 2 pint = ? ½ quart = ? A pint is what part of a quart?

2. 4 quarts = ? ¼ gallon = ? A quart is what part of a gallon?

Change:

3.	**4.**	**5.**
2 qt. to pt.	2 gal. to qt.	12 pt. to gal.
½ qt. to pt.	1½ gal. to qt.	12 qt. to gal.
3½ qt. to pt.	¼ gal. to qt.	10 qt. to gal.

WRITTEN EXERCISE

Change:

1. 7 gal. to qt.
2. 26 gal. to qt.
3. 14 qt. to pint
4. 36 qt. to pt.
5. 19½ qt. to pt.
6. 9 gal. to qt.

Change:

7. 54 pt. to qt.
8. 96 pt. to qt.
9. 184 qt. to gal.
10. 112 qt. to gal.
11. 84 qt. to gal.
12. 192 pt. to gal.

13. How many pints in 9½ gallons?

14. What will a quart of oil cost, if the oil sells for $.40 a gallon?

Measures of Length.

1. How many inches are there in a stick 2 feet 4 inches long?

2. How many feet are there in a room 4 yards 2 feet long?

Table of Length Measure.

12 inches (in.)	= 1 foot (ft.)
3 feet	= 1 yard (yd.)

1. Estimate the length or width of the following: the board ruler; a pencil, your desk, the window, the door, the blackboard, the aisle, the window box, the pictures, the hall, the room.

2. Test your estimate with the ruler, the yard stick, or the tape.

WRITTEN EXERCISE

Change:

1. 9 ft. to in.
2. $3\frac{1}{2}$ ft. to in.
3. 13 ft. to in.
4. 16 yd. to ft.
5. 48 yd. to ft.
6. 9 yd. to in.

Change:

7. 36 in. to ft.
8. 144 in. to ft.
9. 39 ft. to yd.
10. 156 ft. to yd.
11. 180 in. to ft.
12. 324 in. to yd.
13. How many inches in 9 ft. 7 in.?

ORAL PROBLEMS

1. Margaret paid $.11 a quart for peas. What would a peck of peas cost her?

2. Helen put 2 pecks of beans into quart bags. How many bags did she fill?

3. The farmer's chickens eat 2 quarts of corn every day. How long will a peck last?

4. John bought 4 quarts of cherries at $.05 a pint. How much did he pay for them?

5. I sold a peck and 2 quarts of chestnuts at 8 cents a quart. How much did I receive?

6. What will 2 pecks of potatoes cost at one dollar a bushel?

7. Peanuts sell for $2 a bushel. What does a peck sell for?

8. Cranberries sell for $.15 a quart. They cost the grocer 10¢ a quart. How much will he make in selling a peck?

WRITTEN PROBLEMS

1. If peaches are 18¢ a quart, what is the price of 2 pecks?

2. How much would 2 quarts and 1 peck cost?

3. Spinach is 8¢ a quart. How much change will Harry bring home from $2 if he buys 1 peck of spinach?

4. How much will a bushel of tomatoes cost at 14¢ a quart?

5. The grocer sold 118 bushels of potatoes. How many pecks did he sell?

6. We have used 2 quarts of potatoes every day for the last 16 days. How many pecks have we used? How many bushels?

7. If tomatoes cost $1.12 a peck, how much will 4 bushels cost?

Time Measure.

1. Name the days of the week. Name the months of the year.

2. Learn the following:

> Thirty days has September,
> April, June, and November.
> All the rest have thirty-one
> Excepting February alone,
> To which we twenty-eight assign,
> Till leap year gives it twenty-nine.

Table of Time Measure.

60 seconds (sec.)	=1 minute (min.)
60 minutes	=1 hour (hr.)
24 hours	=1 day (da.)
7 days	=1 week (wk.)
4 weeks (about)	=1 month (mo.)
365 days	=1 year (yr.)
366 days	=1 leap year
12 months	=1 year
100 years	=1 century

WRITTEN EXERCISE

1. How many hours in 5 da. 4 hr.?

Change:

2. $7\frac{1}{2}$ yr. to mo.	5. $2\frac{2}{3}$ hr. to min.
3. 13 wk. to da.	6. 9 hr. to min.
4. $4\frac{1}{4}$ da. to hr.	7. $3\frac{1}{2}$ da. to hr.

The Thermometer.

The thermometer is an instrument used to measure **temperature**. The liquid you see in the tube rises when the air becomes warmer, and falls when the air grows cooler. The marks along the tube are called **degrees**. Noting the position of the liquid is called **reading** the thermometer.

1. Find the point marked 0°.

2. Find the freezing point, marked 32°. When the temperature falls to 32°, water freezes. It boils at 212°, which is called the boiling point.

3. How many degrees between zero and the boiling point? Between zero and the freezing point?

4. How many degrees between the freezing point and the boiling point?

5. Read the thermometer now.

6. Keep a record of the temperature for five days, reading the thermometer at the same time every day.

ORAL DRILL EXERCISE

	A	B	C	D
1.	41×5	80×8	60×9	63÷9
2.	9×4+5	3×12+7	2×8+5	70×6
3.	108÷9	100÷10	81÷9	6×6+7
4.	90×4	54÷9	120×3	36÷9
5.	90÷9	41×7	80÷10	31×5
Divide:				
6.	5)55	48÷2	5)45	36÷3
7.	32÷2	3)90	48÷4	9)27
8.	9)180	66÷3	6)18	40÷2
9.	48÷4	4)160	40÷4	7)140
10.	6)36	84÷2	4)28	60÷5
Give results:				
11.	43×2	?×12=84	27+36	24×4
12.	?×9=36	120×5	?×4=48	45+48
13.	91×9	19+42	150×3	?×5=60
14.	54+26	80×4	65+45	52+27
15.	34×2	32×4	?×5=55	33×3
Tell the answers:				
16.	$\frac{1}{3}$ of 15	63+23	$\frac{1}{10}$ of 80	56−37
17.	$\frac{1}{4}$ of 48	72+46	$\frac{1}{8}$ of 56	84−25
18.	$\frac{3}{4}$ of 24	90+54	$\frac{1}{9}$ of 36	19−12
19.	$\frac{1}{5}$ of 35	81+22	$\frac{1}{5}$ of 45	27−18
20.	$\frac{1}{6}$ of 36	46+36	$\frac{2}{3}$ of 30	91−80

GENERAL ORAL PROBLEMS

1. How many pints are there in 2 quarts of milk? In 1 gallon?

2. How many pint glasses can I fill with 2 quarts of water? 1 gallon of water?

3. How many pints in 3 quarts and 1 pint?

4. How many ounces in $\frac{1}{4}$ of a pound?

5. How many in $\frac{1}{8}$ of a pound? In $\frac{3}{4}$ of a pound?

6. A sample box holds $\frac{1}{4}$ of a pound of sugar. How many boxes can I fill with 4 pounds of sugar?

7. How many ounce boxes will be needed to hold 3 pounds of tea?

8. Jerry and Louis together weigh 120 pounds. Jerry weighs 70 pounds. How much does Louis weigh?

9. Mabel has saved $.28 to buy a book that costs $.50. How much more must she save?

10. John gave the baker a dollar to pay his mother's bill of 72 cents. How much change will he receive?

11. Ethel took out $.32 from her bank. Now she has $.64 left in it. How much did she have at first?

12. If oranges are 4¢ each, how many can I buy for 48 cents?

13. Ethel has 12 cents. How many peaches can she buy at 3 cents each?

14. If Helen has $48 in her bank and takes out $\frac{1}{4}$ of it, how much does she leave in the bank?

GENERAL WRITTEN PROBLEMS

1. A quarter-pound package of chocolate costs 21 cents. Find the cost of three pounds at this rate.

2. I bought 22 yards of cloth at $.37 a yard. What did it cost me?

3. The grocer sold 94 quarts of potatoes. He has left 69 quarts. How many quarts did he have at first?

4. His rent last month was $118.50; his gas bill $22.14; his feed bill $27.42. Find what he paid for these three items.

5. The grocer boy earns $6.50 a week. How much will he earn in 12 weeks?

6. How many 8-ounce packages of chocolate can be made from 41 pounds of chocolate?

7. How many pounds of butter in 22 tubs, each containing 65 pounds?

8. How many gallons in 153 quarts of oil? How many quarts will be left over?

9. How many yards in 53 feet of wire? How many feet are left over?

10. A driver is paid $18.42 for 6 days' work. Find what he gets a day.

11. Find what the driver will earn in 19 days. In 24 weeks.

12. If he spends an average of $14.00 a week, how much will he spend in 24 weeks?

ORAL REVIEW PROBLEMS

1. There were 40 girls in the relay race. $\frac{1}{8}$ dropped out. How many dropped out?

2. 8 men carried 72 chairs for Field Day. How many did each carry if they divided the number of chairs equally?

3. The music and the programs together cost $15. The programs cost $3. What did the music cost?

4. The band played 7 pieces every hour. How many did they play in 6 hours? How many times 2 is 6?

5. Harry had $.80 after selling papers. $\frac{1}{4}$ of this was profit. How much was profit?

6. The janitor cleaned $\frac{2}{3}$ of our 15 window panes. How many did he clean?

7. It takes him 3 minutes to clean one pane. How long will it take to clean the 15?

8. There are 48 boys in our class. $\frac{1}{8}$ were not promoted. How many were not promoted?

9. A farmer had 36 chickens. He sold $\frac{1}{4}$ of them. How many did he keep?

10. If John pays 3 cents for 5 newspapers, how much money must he have for 25 newspapers?

11. If he sells 50 newspapers daily at 1 cent a piece, how much will he earn in one day?

12. How many 3's in 18? If skates are $3 a pair, how many pairs can I buy for $18?

WRITTEN REVIEW PROBLEMS

1. Our class had $5.16 in the treasury. Last week we added $2.87. How much did we then have?

2. We paid the carfare to the museum and back for 38 pupils at 10 cents each. What did this cost? How much have we left in our treasury?

3. The grocer bought a wagon for $42.25. He sold it for $6.50 less. What did he get for it?

4. The iceman sells an average of $9.40 worth of ice every day. How much does he sell in a week of 6 days?

5. Each day his ice costs him $6.30. How much does he pay for ice every week?

6. To-day he has $164.30 in the bank. Last week he drew out $10.50 to buy a coat. How much had he in the bank at first?

7. This summer he sells about 90 pieces of ice a day. Last summer he sold $\frac{1}{3}$ as much more. How much did he sell daily last summer?

8. Harry bought a pair of skates for $1.25 and a baseball mitt for $.69. How much change will he bring home from a five-dollar bill?

9. Each family in Helen's house pays $22 a month rent. There are 12 families. How much rent does the owner of the house get every month?

10. $\frac{3}{4}$ of the families pay their rent on the first day of the month. How many do not pay on the first day?

11. It cost us $12.50 to move last month; our rent was $22; and our gas bill $3.37. Find the total of these expenses.

12. If our rent was $22 a month and our grocery bill about the same as our rent, what do the two expenses amount to in a year?

13. A phonograph with records cost $33.75. The phonograph cost $24. Find what the records cost.

14. A better phonograph with records would cost $45. How much more would this be than the first?

15. I sold the phonograph that cost $33.75 and made $9.50 on it. How much did I receive for it?

16. Mr. Adams drew out $100 from the bank to buy furniture. He bought a dresser for $21.25, a bedstead for $39.50, and 6 pictures for $19.20. How much money had he left?

17. His wife spent $9.50 for a hat and three times as much for a new suit. How much did she spend?

18. Mr. Adams left the house at 9.30 A. M. and returned at 12 M. Mrs. Adams left at the same time but was gone twice as long. When did she get home?

19. One quarter of a yard of silk costs $.42. What will $\frac{1}{2}$ a yard cost?

20. A man earns $160 a month and spends $125 a month. How much does he save in 8 months?

21. Find the cost of 24 armchairs at $48 each.

22. A dealer paid $256 for 8 electric lamps. Find the cost of each lamp.

23. There were 357 pupils in 10 classes. $\frac{1}{7}$ of them were not promoted. How many were not promoted?

24. 8460 books were divided equally among 5 schools. How many books did each school receive?

25. Three men bought an automobile for $2274. If their shares are equal, find what each share is worth.

26. Tony's grocery bill for one week was $4.87; his butcher's bill was $3.65; his baker's bill $1.20; and his laundry bill $1.06. Find what the four bills amounted to.

27. His rent is $22.50 a month. This is $7.40 less than his rent before he moved. How much did he pay then?

28. Last week he sold 7 doz. apples and 6 times as many peaches. How many peaches did he sell?

29. Last week he lost $2.34 on peaches for which he paid $11.50. How much did he receive for them?

30. What will 9 horses cost at $345 each?

31. A butcher's ice bill for one month was $56.90. The next month it was $91.26. How much more did he pay for ice the second month?

32. From a load of 28 pecks of apples, how many quart pails can a pedlar fill?

33. The grocer wants to buy 13 gas lamps for his store. They cost $12 apiece. What will he pay?

34. Oil costs him $.50 a gallon. How many pint bottles can he fill if he buys $1.50 worth of oil?

TABLES FOR REFERENCE

Length

12 inches (in.)	=1 foot (ft.)
3 feet	=1 yard (yd.)

Liquid Measure

2 pints (pt.)	=1 quart (qt.)
4 quarts	=1 gallon (gal.)

Dry Measure

2 pints (pt.)	=1 quart (qt.)
8 quarts	=1 peck (pk.)
4 pecks	=1 bushel (bu.)

Time Measure

60 seconds (sec.)	=1 minute (min.)
60 minutes	=1 hour (hr.)
24 hours	=1 day (da.)
7 days	=1 week (wk.)
4 weeks (about)	=1 month (mo.)
365 days	=1 common year (yr.)
366 days	=1 leap year
12 months	=1 year
100 years	=1 century

Weight Measure

16 ounces (oz.)	=1 pound (lb.)

United States Money

10 cents	=1 dime (d.)
10 dimes	=1 dollar ($)
100 cents	=1 dollar ($)

MULTIPLICATION TABLES

2× 1= 2	3× 1= 3	4× 1= 4	5× 1= 5
2× 2= 4	3× 2= 6	4× 2= 8	5× 2=10
2× 3= 6	3× 3= 9	4× 3=12	5× 3=15
2× 4= 8	3× 4=12	4× 4=16	5× 4=20
2× 5=10	3× 5=15	4× 5=20	5× 5=25
2× 6=12	3× 6=18	4× 6=24	5× 6=30
2× 7=14	3× 7=21	4× 7=28	5× 7=35
2× 8=16	3× 8=24	4× 8=32	5× 8=40
2× 9=18	3× 9=27	4× 9=36	5× 9=45
2×10=20	3×10=30	4×10=40	5×10=50
2×11=22	3×11=33	4×11=44	5×11=55
2×12=24	3×12=36	4×12=48	5×12=60

6× 1= 6	7× 1= 7	8× 1= 8	9× 1= 9
6× 2=12	7× 2=14	8× 2=16	9× 2= 18
6× 3=18	7× 3=21	8× 3=24	9× 3= 27
6× 4=24	7× 4=28	8× 4=32	9× 4= 36
6× 5=30	7× 5=35	8× 5=40	9× 5= 45
6× 6=36	7× 6=42	8× 6=48	9× 6= 54
6× 7=42	7× 7=49	8× 7=56	9× 7= 63
6× 8=48	7× 8=56	8× 8=64	9× 8= 72
6× 9=54	7× 9=63	8× 9=72	9× 9= 81
6×10=60	7×10=70	8×10=80	9×10= 90
6×11=66	7×11=77	8×11=88	9×11= 99
6×12=72	7×12=84	8×12=96	9×12=108

ROMAN NUMERALS

I = 1	XII = 12	XXIII = 23
II = 2	XIII = 13	XXIV = 24
III = 3	XIV = 14	XXV = 25
IV = 4	XV = 15	XXVI = 26
V = 5	XVI = 16	XXVII = 27
VI = 6	XVII = 17	XXVIII = 28
VII = 7	XVIII = 18	XXIX = 29
VIII = 8	XIX = 19	XXX = 30
IX = 9	XX = 20	XL = 40
X = 10	XXI = 21	L = 50
XI = 11	XXII = 22	

CONTENTS

FOURTH YEAR BOOK

FIRST HALF: GRADE 4 A

SECOND HALF: GRADE 4 B

ARITHMETIC BY GRADES

FOURTH YEAR BOOK

FIRST HALF: GRADE 4A

I. READING AND WRITING NUMBERS

ORAL EXERCISE

1. Read:

4,217	7,613	8,255	10,250	25,000
311	8,000	6,066	15,900	36,240

2. Tell which figures stand for *units, tens, hundreds, thousands,* and *ten thousands* in these numbers:

9 17 28 70 411 5,617 9,109 11,228

3. Read these numbers:

18,996	42,638	13,640	57,001
35,723	28,000	45,961	50,019
26,026	37,321	19,371	49,940

Remember that we do not use "and" in reading these numbers. We read, "eighteen thousand nine hundred ninety-six."

4. Read:

$46	$572	$881	$15,247	$55,910
$12	$148	$4,360	$34,950	$18,463

WRITTEN EXERCISE

Write in figures:

1. Five hundred nine.
2. Twelve thousand.
3. Sixteen thousand.
4. Twenty-two thousand.
5. Nineteen thousand, nine hundred seven.
6. Thirty-one thousand, three hundred.
7. Twenty-nine thousand, one hundred three.

ORAL EXERCISE

1. Count by 1000's from 40,000 to 100,000.
2. Read:

57,314	71,380	85,264	98,881
60,000	79,900	88,365	80,000
30,000	40,000	70,000	100,000
44,117	69,019	53,003	87,027

ORAL EXERCISE: ADVANCED WORK

1. Read:

300,000	562,370	825,221	910,000
421,000	731,060	840,008	990,900
255,100	647,000	778,000	345,050

2. Count by 10,000's to 100,000. By 100,000's to 900,000. From 510,000 to 610,000.

WRITTEN EXERCISE: ADVANCED WORK

Write in figures:

1. Two hundred thousand.
2. Five hundred thousand.
3. One hundred sixty-five thousand, two hundred.
4. Four hundred seven thousand, eighteen.
5. Seven hundred seventy thousand, three hundred.
6. Six hundred fifty-one thousand, nine hundred six.

United States Money.

1. Read: $1.25 $17.36 $163.15 $742.19
$3.39 $28.44 $478.68 $1462.45

When we write dollars and cents, we use the **decimal point** to separate the dollars from the cents. $163.15 is read: "one hundred sixty-three dollars *and* fifteen cents."

We may read a number like $1247 in two ways: namely, "twelve hundred forty-seven dollars," and "one thousand, two hundred forty-seven dollars."

When the number of cents is less than 10, we write 0 in tens' place, thus:

$5.06 Five dollars and six cents

2. Read: $3.02 $4.04 $36.07 $48.05 $17.09
$7.03 $9.00 $124.01 $563.06 $984.08

We write: Seven cents $.07
Thirty-two cents $.32

3. Read: $.14 $.03 $.39 $.75
$.72 $.59 $.09 $.50
$126.02 $58.31 $599.83 $675.02

WRITTEN EXERCISE

Write in figures:

1. Forty-six thousand.
2. Sixty-nine thousand.
3. Eleven thousand five.
4. Ninety-nine thousand.
5. Seventy-five thousand, one hundred eight.
6. Thirty-nine cents.
7. Fifty-four dollars.
8. Eleven dollars and twenty-five cents.
9. Two hundred eight dollars and sixty-three cents.

Write in words:

10.	$25.39	68,412	$111.16	49,003	$336.57
11.	$169.15	75,000	$8,000	97,107	$7,116.20
12.	$4.16	48,700	$5,250.10	33,396	$4,004.08

Roman Numerals. Notice how the pages in the preface of this book are numbered. The Romans used this method of numbering; that is, they used letters instead of figures.

I=1	VI= 6	X=10	LX= 60
II=1+1	VII= 7	XX=20	LXX= 70
III=3	VIII= 8	XXX=30	LXXX= 80
IV=5−1	IX=10−1	XL=50−10	XC=100−10
V=5	X=10	L=50	C=100

Notice that when a letter is repeated the value is repeated: XX=X+X.

When a letter is placed after one of greater value, its value is added to that of the greater: XI=X+I.

When a letter is placed before one of greater value, its value is taken away from that of the greater: IX=X−I.

1. Read:

IV	IX	XXV	XXXIV	L	XL	LX
XLVI	XLI	XXVII	LXXV	C	LVI	XC

Roman numbers are used on the faces of clocks and watches; and for numbering the chapters, and sometimes the pages, of books. They are found also on some coins, medals, and monuments.

2. Tell which chapters of a book the following Roman numerals designate:

XI	XVII	XXIII	XXIX	XXXIII	XXXVIII
XLII	XLIX	LII	VI	LXXVI	XLVIII

3. Write in Roman numbers:

5	10	15	20	25	35	45	55
70	85	100	11	24	39	46	68
89	90	91	92	93	97	99	200

Roman Numerals: Advanced Work. The letter C stands for 100. The letter D for 500. The letter M for 1000. The Romans used seven letters.

Letters:	I	V	X	L	C	D	M
Values:	1	5	10	50	100	500	1000

The hundreds are:

C	CC	CCC	CD	D	DC	DCC	DCCC	CM	M
100	200	300	400	500	600	700	800	900	1000

1. Read:

CL	DCXX	CDXX	CCCLX	DLXV	M
DLXIV	CLXXI	DCXLV	DCVI	CMLIX	CMXI
CXVII	CCXXX	CIV	DCCVI	CLIII	DCIX

2. Write in Roman numbers:

105	503	880	900	1000
430	346	725	812	231
621	783	411	555	640
1916	1917	1918	1919	1920

II. COUNTING

ORAL EXERCISE

I. 1. Count by 5's from 0 to 60.
2. Count by 2's from 14 to 48.
3. Count by 3's from 7 to 58.
4. Count by 4's from 2 to 50.
5. Count by 10's from 10 to 120.
6. Count by 6's from 20 to 92.
7. Count by 7's from 2 to 86.

II. 1. Beginning with 11, count by 4's to 59.
2. Beginning with 20, count by 10's to 120.
3. Beginning with 10, count by 5's to 80.
4. Beginning with 6, count by 6's to 84.

III. 1. Count backward by 5's from 58 to 3.
2. Count backward by 4's from 55 to 11.
3. Count backward by 3's from 56 to 11.

IV. 1. Count by 1000's from 26,000 to 40,000.
2. Count by 2000's from 41,000 to 59,000.

WRITTEN EXERCISE

1. Write the 100's from 1500 to 2500.
2. Write the 1000's from 19,000 to 33,000.
3. Write the 6's from 7 to 73.
4. Write the 7's from 10 to 80.
5. Write the 8's from 15 to 87.
6. Write the 9's from 1 to 108.

ORAL EXERCISE

I. 1. Beginning with 9, count to 89 by 8's.
2. Beginning with 12, count to 102 by 9's.
3. Beginning with 4, count to 85 by 9's.
4. Beginning with 5, count to 75 by 7's.
5. Count backward by 4's from 95 to 3.
6. Count backward by 7's from 65 to 2.
7. Count backward by 3's from 61 to 1.
8. Count backward by 8's from 74 to 2.
9. Count by 1000's from 87,000 to 100,000.

II. 1. Beginning with 3, count by 9's to 75.
2. Beginning with 4, count by 8's to 68.
3. Beginning with 5, count by 7's to 82.
4. Beginning with 6, count by 6's to 72.
5. Beginning with 7, count by 5's to 62.
6. Beginning with 8, count by 4's to 48.

WRITTEN EXERCISE

1. Write the 6's from 53 to 5.
2. Write the 7's from 60 to 4.
3. Write the 100's from 7300 to 6400.
4. Write the 1000's from 97,000 to 85,000.
5. Write the 9's from 89 to 8.
6. Write the 10's from 920 to 1050.
7. Write the 4's from 37 to 81.
8. Write the 200's from 8800 to 5800.
9. Write the 600's from 10,000 to 2800.

III. ADDITION

ORAL EXERCISE

Give these sums quickly:

1.	$.60	$.50	$.70	$.80	$.40	$.30	$.10
	.30	.10	.20	.10	.40	.20	.30
	.10	.20	.10	.10	.10	.20	.40

2.	20¢	70	50¢	20	20¢	50	20
	30¢	20	20¢	40	30¢	30	20
	40¢	10	10¢	50	20¢	10	20

Terms in Addition. In addition the numbers to be added are called the **addends.** The answer is called the **sum**, the **total**, or the **amount**.

We **check** or **prove** our work in addition by adding in the opposite direction, first *down*, and then *up*.

ORAL DRILL

	A	*B*	*C*	*D*	*E*	*F*	*G*	*H*	*I*	*J*
a	7	9	8	9	7	4	4	8	9	4
b	3	2	9	7	2	7	2	5	6	7
c	6	8	5	8	8	9	8	5	9	9
d	9	7	7	6	9	8	4	6	9	5
e	8	8	8	8	5	5	1	8	4	5
f	1	1	1	6	8	9	6	9	8	8
g	4	9	4	3	1	8	3	5	1	4
h	5	8	7	1	9	5	8	6	3	6

Look down and then up each column and see how many groups of numbers you can find that will make **ten**, *such as:*

5	6	1	6	4	1	2	9
5	4	2	3	3	4	8	1
		7	1	3	5		

Where possible, make use of these groups of 10, *since this method helps you to add the columns with greater speed.*

In the oral drill, add each column down and then up.

WRITTEN EXERCISE

Add and check your results:

1.	**2.**	**3.**	**4.**	**5.**
1264	439	$79.74	1200	2246
1397	2067	1.29	896	843
2533	1073	6.25	1115	1509
1028	832	14.50	674	1774
1062	346	3.98	3048	1492

6.	**7.**	**8.**	**9.**	**10.**
4974	8296	7672	9413	6784
3836	5476	4873	8478	7392
5089	7284	9283	7645	7897
7446	9095	6789	5994	5647

11.	**12.**	**13.**	**14.**	**15.**
682	2075	$62.81	4078	$21.29
705	1008	20.02	5634	31.84
493	4163	73.81	1257	76.89
539	2084	10.79	7284	2.34
722	3078	25.67	2073	50.65
846	161	45.83	4592	43.24

Add and check:

16.	17.	18.	19.	20.
8025	9079	$ 7.67	8148	343
364	5000	41.03	729	1300
1297	4263	.39	6134	864
4524	234	.69	139	547
762	3009	3.25	5009	5843
1893	6789	14.49	7017	982

ORAL EXERCISE

1. Add 43 and 35:

43 35 78	Here it is easier to think of 35+40=75 and then 75+3=78. Begin at the left and add the tens first.

2. Add:

42	21	63	51	47	85	34	79
24	12	36	15	74	54	43	96

3. In the drill chart on page 8, add each row across from left to right. Add each from right to left.

Add rapidly:

4.	14	64	74	54	34	44	21	62	75
	75	14	23	32	46	56	68	27	14
5.	55	63	18	34	47	52	11	16	27
	43	26	71	45	32	27	78	52	71
6.	22	33	44	55	66	77	85	15	47
	39	48	57	26	32	14	12	74	52

ORAL EXERCISE

Give these sums:

1. 19	22	24	48	35	43	24	16
44	57	68	45	65	46	57	36

2. 29	32	45	22	14	13	28	35
58	49	55	62	77	69	47	37

3. $.50	$.80	$.60	$.40	$.50	$.90	$.80
.40	.30	.20	.10	.20	.30	.40
.30	.20	.40	.60	.40	.20	.30

WRITTEN EXERCISE

Add and check:

1.	2.	3.	4.	5.
4,265	8,038	21,160	$ 18.69	$ 3.25
387	503	8,345	3.25	14.19
423	1,269	9,009	101.22	16.78
1,696	778	1,416	14.39	54.36
505	1,492	768	5.25	96.10
415	338	537	18.50	5.25
2,306	1,887	4,334	5.10	12.39

6.	7.	8.	9.	10.
8,275	4,134	12,038	19,876	16,756
549	5,639	786	6,040	2,348
6,076	786	11,386	4,501	8,889
5,125	16,440	4,884	4,975	28,041
438	8,192	7,175	28,463	7,000
3,346	563	16,380	8,674	36,957
1,112	14,143	130	37,375	7,456

ORAL EXERCISE

Give answers:

1.	37	43	82	57	12	38	56	63	71
	62	46	18	42	77	51	43	27	24
2.	81	14	36	73	21	42	37	65	83
	18	74	43	25	78	37	52	34	15

3. In the chart on page 8, add from right to left row *a*; row *b*; *c*; *d*. Now add from left to right.

ORAL EXERCISE

Add:

1. 40	\$.50	60¢	70	30	\$.40	60	50
30	.20	20¢	50	40	.20	30	20
30	.40	40¢	20	60	.50	80	50
2. 60	\$.80	70¢	80	60	\$.40	50	90
50	.50	20¢	20	50	.80	40	60
40	.20	50¢	60	20	.50	30	20

3. In the chart on page 8, add from right to left row *e*; row *f*; *g*; *h*; *i*. Check by adding from left to right.

Give these sums:

4.	34	16	44	25	19	37	54	64
	27	38	27	44	53	46	25	28
5.	45	63	74	18	35	44	65	55
	34	29	18	52	47	49	18	26

WRITTEN EXERCISE

Add and check:

1. 16,755	2. $943.32	3. 18,077	4. 6,756
4,287	150.08	2,824	22,834
8,800	34.45	46,520	2,803
43,579	45.00	6,482	6,903
11,965	37.00	12,582	14,735
1,236	80.88	8,457	8,569
6,000	62.50	4,400	12,113
4,811	148.57	16,253	1,927

5. 16,402	6. 46,053	7. 26,756	8. $312.34
738	4,028	6,957	90.31
1,346	5,840	7,187	367.34
937	23,184	8,228	553.93
12,002	8,034	10,371	19.51
8,246	5,308	4,252	150.12
986	13,644	3,876	127.34
1,124	2,136	17,532	83.74

9. 4,508	10. 2,209	11. $45.08	12. 19,345
35,702	8,216	488.07	1,234
19,646	15,702	206.46	6,734
1,927	5,393	481.20	21,260
4,812	2,504	25.82	1,982
28,815	24,502	138.47	32,200
2,700	1,708	35.56	8,815
1,551	462	5.04	705

WRITTEN REVIEW PROBLEMS

1. A cashier took in $24.68 on one day, $34.16 on another, $21.05 on a third day, and $54.73 on a fourth. How much money did he receive in all?

2. A passenger elevator carried 439 people the first hour, 388 the second, 412 the third, 501 the fourth, and 447 the fifth. How many passengers were carried in the five hours?

3. A man saved $12.68 in July, $17.23 in August, $24.12 in September, $17.50 in October, and $12.58 in November. How much money did he save in all?

4. A shoe dealer had 149 pairs of boys' shoes, 237 pairs of girls' shoes, 543 pairs of women's shoes, and 743 pairs of men's shoes. Find the amount of his stock.

5. How many pupils are there in a school district if the first school has 1114, the second 2538, the third 896, the fourth 1243, the fifth 1786, and the sixth 2469?

6. A newsboy sold 384 papers the first week, 249 the second, 324 the third, and 318 in the fourth week. How many did he sell during the month?

7. A merchant put $4312 in the bank one week, $3187.02 the second, $2916 the third, and $5101.82 the fourth. Find the total amount of his deposits.

8. A postman collected 713 letters the first day, 832 the second, 689 the third, 593 the fourth, and 781 the fifth. How many letters did he collect in all?

9. On Monday a milkman delivered 420 quarts of milk, on Tuesday he delivered 396 qt., Wednesday 416, Thursday 404, Friday 383. How many did he deliver in the five days?

10. A restaurant manager checked the following amounts of rice: 3815 lb., 2156 lb., 3404 lb., 4112 lb., and 3897 lb. How many pounds did he receive?

11. On Sunday 847 people visited the park in automobiles, 1122 came by car, and 623 walked. How many people came in all?

12. The owner of a four-story house received in rents: $324.50 from the first floor, $415.40 from the second, $398 from the third, and $347.25 from the top floor. Find how much rent he received from the house.

13. How many books are there in the supply closet if the first shelf holds 234 books, the second 315 books, the third 503 books, the fourth 89 books, the fifth 377 books, and the top shelf 116 books?

14. Mr. Eltinge spent $.98 for a cap, $7.85 for two pairs of shoes, $15.25 for an overcoat, and $5.20 for shirts. How much did he spend in all?

15. John rode in his father's car 48 mi. on Sunday, 113 mi. on Monday, 79 mi. on Wednesday, and 129 mi. on Friday. How far did he ride during the four days?

16. Last year Mrs. Allen deposited in a savings bank the following amounts: $55.28 in January, $18.46 in April, $12.94 in June, and $48.93 in August. How much did she deposit in all?

IV. SUBTRACTION

ORAL EXERCISE

1. From:

From:	19	18	17	19	35	32	65	45
take:	12	9	11	8	15	18	20	22

2. In the second line of Ex. **1**, subtract each number from 100.

Subtract rapidly:

3.	56	67	77	78	85	89	94	98
	−25	−32	−44	−45	−64	−27	−33	−55
4.	74	47	53	80	70	60	50	40
	−30	−12	−13	−35	−25	−15	−15	−15
5.	90	60	80	40	70	90	50	20
	−74	−24	−54	−14	−34	−44	−24	−14
6.	15	65	85	35	55	95	45	75
	−6	−26	−56	−26	−36	−66	−26	−46

Terms in Subtraction. **The sign of subtraction** is −, which is read *minus.*

The **minuend** is the number from which another number is taken.

The **subtrahend** is the number to be taken away.

The result in subtraction is called the **difference,** or the **remainder.**

Subtracting Numbers of Three Figures.

From 962 take 576:

962	*minuend*	6+**6**=12.	Write 6; add 1 to 7.
576	*subtrahend*	8+**8**=16.	Write 8; add 1 to 5.
386	*difference*	6+**3**=9.	Write 3.

To prove the answer we add the *difference* and the *subtrahend* without rewriting them. The sum should equal the *minuend*.

576+386=962.

WRITTEN EXERCISE

Subtract and prove:

1. 314	**2.** 522	**3.** 734	**4.** 641	**5.** 817	**6.** 602
285	438	649	547	709	385

ORAL EXERCISE

Subtract:

1.	14	26	33	78	55	47	94	83
	−9	−13	−20	−52	−40	−26	−54	−51
2.	34	46	32	41	37	28	24	21
	−18	−24	−15	−35	−18	−19	−15	−14
3.	68	53	85	60	46	51	40	31
	−32	−47	−36	−39	−31	−35	−19	−18
4.	67	97	17	47	77	27	57	87
	−38	−28	−8	−18	−38	−18	−28	−48

Subtracting Numbers of Four Figures.

Subtract 4268 from 7103:

7103	8+**5**=13.	Write 5; add 1 to 6.
4268	7+**3**=10.	Write 3; add 1 to 2.
	3+**8**=11.	Write 8; add 1 to 4.
2835	5+**2**= 7.	Write 2.

Proof: 4268+2835=7103.

WRITTEN EXERCISE

Subtract and prove:

1. 8307 2961	**2.** 8000 2168	**3.** 7500 3962	**4.** 15000 11448
5. 9875 6541	**6.** 7035 5296	**7.** 3806 1987	**8.** 3000 1189
9. 65111 42225	**10.** 9000 8296	**11.** 13611 2815	**12.** 8412 6574
13. 2407 3625	**14.** 34711 9628	**15.** 18432 7685	**16.** 22024 12356
17. 28871 10980	**18.** 24346 877	**19.** 13000 216	**20.** 12723 5875
21. 3000 298	**22.** 4400 3528	**23.** 5000 179	**24.** 11410 3847

ORAL EXERCISE

Subtract:

1.	24	71	26	42	80	63	94	33
	17	38	19	33	69	37	65	22
2.	85	54	72	84	96	62	74	80
	67	45	48	59	64	38	49	69

Subtract each of the following numbers from 100:

3.	12	19	40	32	84	60	13	48
4.	66	23	69	44	17	99	91	62
5.	20	94	56	55	73	81	35	76
6.	59	26	67	33	64	72	89	53

7. You give the butcher a two-dollar bill in payment for a $1.45 purchase. How much change should you receive?

While giving you change, if he hands you a 5-cent piece and a half-dollar, the butcher will probably say, "$1.45, $1.50, $2.00."

8. Make the change for these purchases, assuming that you have given the clerk the amount at the left:

	a	*b*	*c*	*d*	*e*	*f*
$1.00	$ .15	$ .25	$ .35	$ .39	$ .49	$ .55
$2.00	$1.15	$1.40	$1.65	$1.73	$1.55	$1.61
$5.00	$3.50	$4.25	$2.70	$2.15	$3.20	$4.82
$3.00	$2.25	$2.39	$2.10	$2.56	$2.75	$2.85
$10.00	$9.25	$6.46	$7.15	$8.66	$9.70	$7.43

Subtracting Dollars and Cents.

Subtract $288.25 from $415.12:

	5+**7**=12.	Write 7; add 1 to 2.
$415.12	3+**8**=11.	Write 8; add 1 to 8.
288.25	9+**6**=15.	Write 6; add 1 to 8.
$126.87	9+**2**=11.	Write 2; add 1 to 2.
	3+**1**= 4.	Write 1.

Proof: $288.25+126.87=415.12.

In subtracting dollars and cents, we write the decimal points in a column. Write the dollar sign before the minuend and the remainder.

WRITTEN EXERCISE

Subtract and prove your results:

1. $86.45 72.24	**2.** $24.36 20.47	**3.** $47.91 37.96	**4.** $39.93 19.87
5. $62.12 59.58	**6.** $74.15 33.46	**7.** $85.65 37.76	**8.** $21.15 9.29
9. $212.24 162.59	**10.** $314.16 247.89	**11.** $478.51 173.65	**12.** $137.19 39.99
13. 1538 1449	**14.** 9482 7596	**15.** 12362 8485	**16.** 18801 14965
17. 5423 3679	**18.** 7220 4829	**19.** 10970 8282	**20.** 15400 7952

21. $148.63 139.79	**22.** 22266 21879	**23.** $154.62 129.85	**24.** 28810 4961
25. $814.20 625.37	**26.** 32100 17585	**27.** $1622.15 486.29	**28.** 46342 9861

WRITTEN REVIEW PROBLEMS

1. The attendance of our school on Monday was 2411. On Tuesday it was 2389. Find the difference in the attendance during the two days.

2. An ice dealer received 9461 lb. of ice and sold 7583 lb. How many lb. had he left?

3. Out of an army of 18,290 men, 1175 were killed and 1342 were wounded. How many men were left?

4. A dealer sold a house valued at $12,250 for $15,125. How much money did he gain?

5. A merchant had $4165 in the bank. He put in $2850 more and then he drew out $3110. How much money has he in the bank now?

6. The distance between two cities is 1872 miles. A train has covered 798 miles of this distance. How far has it still to go?

7. A school hall seats 1000 people. 216 seats are filled by visitors and 712 are filled by pupils. How many seats are vacant?

8. A woman bought a mattress for $5.25 and a rug for $9.50. How much change should she get from $20?

9. 874 pupils attended an athletic meet of three schools. The first school sent 268; and the second, 509. How many did the third school send?

10. There are 5280 feet in one mile. How much less than 1 mile is a race course of 4375 feet?

11. A girl bought butter for \$.72, tea for \$.48, crackers for \$.39, and flour for \$1.20. How much change did she receive from \$10?

12. Take the sum of 2247 and 1995 from the sum of 3653 and 889.

13. 1278 persons attended a theatre on Thursday night, 1159 on Friday night, and 1643 on Saturday night. How many attended during the three nights?

14. A farmer received \$487.16 for his apples, \$290.63 for his potatoes, and \$306.85 for lumber. How much did he receive for all?

15. A man drew from his bank \$1085 to purchase an automobile. He then had \$2216.34 left in the bank. How much money had he at first?

16. Subtract three thousand six hundred thirty-seven from five thousand eleven.

17. A business man had \$487.13 in the bank. If he added \$56.05 to his bank account and then drew out \$395.67, how much did he leave in the bank?

18. How much is gained by selling for \$8725 a farm that cost \$5250?

19. From 5984 take the sum of 1873 and 609.

ORAL DRILL EXERCISE

	A		*B*		*C*		*D*	
Add the following:								
1.	230 80	420 90	580 30	370 40	640 70	720 80	920 90	840 70
2.	\$4.50 .90		\$3.20 .80		\$7.20 .90		\$9.10 .80	
3.	50 20 10	60 90 20	40 30 60	60 60 30	50 70 40	40 80 50	70 40 80	30 20 70
State results:								
4.	6×60		9×50		12×100		7×80	
5.	7×70		10×60		10×90		6×70	
6.	8×60		11×50		9×100		5×90	
7.	40)480		70)840		90)360		50)150	
8.	50)600		80)720		60)480		30)210	
9.	35 −19	77 −58	86 −37	95 −48	46 −17	54 −35	68 −28	63 −37
10.	6×81		8×91		9×61		7×31	
11.	7×91		9×71		8×81		6×41	
12.	$\frac{1}{2}$ of 128		$\frac{1}{3}$ of 360		$\frac{1}{5}$ of 350		$\frac{1}{8}$ of 648	
13.	$\frac{1}{3}$ of 240		$\frac{2}{3}$ of 120		$\frac{1}{10}$ of 770		$\frac{1}{7}$ of 147	
14.	$\frac{1}{4}$ of 480		$\frac{1}{5}$ of 350		$\frac{3}{4}$ of 1200		$\frac{1}{6}$ of 366	
15.	$\frac{3}{4}$ of 240		$\frac{1}{6}$ of 420		$\frac{2}{3}$ of 120		$\frac{1}{8}$ of 488	

GENERAL ORAL PROBLEMS

1. If the milkman delivered 38 bottles of milk and has 38 bottles left on his wagon, how many had he at first?

2. He sold 32 bottles. 17 of them were quart bottles. How many were pint bottles?

3. He sold a dozen bottles of cream at $.12 each. How much money did he collect for all of them?

4. The wagon traveled 4 miles a day for 3 days, and 5 miles a day for 2 days. How many miles were traveled during the 5 days?

5. The milkman gave 63 bottles of milk to 21 families. Find the average number given to each family.

6. At one house he sold 9 bottles of milk at 9¢ each. How much change did he give for a two-dollar bill?

7. If 4 oranges cost 15¢, what will a dozen cost?

8. Helen bought 2 pounds of candy at 30¢ a pound, and 2 cakes at 15¢ each. How much money did she spend?

9. How much change will she have from a five-dollar bill?

10. We have 60 children in the glee club; 46 are girls. How many boys are members?

11. If 6 basket balls cost $24, how much will 9 cost?

12. The baker has 18 loaves of rye bread and 3 times as many loaves of white bread. How many more loaves of white bread has he than of rye?

13. Harry had $.92. How much had he left after spending $.29?

14. Gertrude bought 3 yd. of ribbon at 9¢ a yd., and a pair of scissors for $.25. How much did she pay the clerk?

15. Potatoes are 12¢ a quart. How much must one pay for a peck?

16. How much shall I have left of $500 after paying for 4 cutting machines at $115 each?

17. A man earns $3.50 a day. How much can he earn in 10 days?

18. A dealer bought 34 cows at $30 each. How much did he pay for them?

19. If an automobile travels 20 miles an hour, how long will it require to go 440 miles?

20. How many days are there in March, April, and May?

GENERAL WRITTEN PROBLEMS

1. A dealer has 14 pianos in his store. They cost $525 each. How much did he pay for all of them?

2. Gertrude bought 6 pounds of coffee. How many ounces did she buy?

3. The baker bought 4 doz. boxes of cherries at 14¢ a box. What did they cost?

4. In one order a dealer sold $46.25 worth of chairs, $11.09 worth of pictures, a bed worth $65.50, and a sofa worth $22.25. Find the amount of the bill.

5. Furniture that cost $1872.15 he sold for $2250. How much did he gain?

6. How many dozen cans of corn are there in 1104 cans?

7. If handkerchiefs are sold at 8 for a dollar, how much will 6 cost?

8. Sam's father earns $4.20 a day. Last month he worked 18 whole days, and one half day. How much money did he receive?

9. If 2 feet of cloth costs 86¢, what will a yard cost?

10. The quotient is 250. The divisor is 64. Find the dividend.

11. On each trip a trolley car carries 64 persons at 5¢ per person. How much money does the company receive if the car makes 18 trips a day?

12. From $\frac{2}{3}$ of 882 take $\frac{3}{4}$ of 592.

13. How much will a grocer receive for apples if he sells 128 pk. in quart boxes at 12¢ a quart?

14. A school spent $846 for new geographies. If the books cost $3 each, how many books were ordered?

15. A dealer drew $3465 from his bank and used $\frac{2}{3}$ of the money in buying boys' hats. How much did the hats cost?

16. A girl bought half a dozen ties at 49¢ each. How much change should she receive from $5?

17. A school has 2448 pupils on the roll. If $\frac{1}{9}$ of them are absent, how many are in the school?

18. From $\frac{1}{6}$ of 942 take $\frac{1}{8}$ of 736.

V. MULTIPLICATION

ORAL EXERCISE

Give the products at sight:

1.	2.	3.	4.	5.	6.
7×3	4×8	3×9	2×11	4×7	11×3
8×5	12×5	10×5	4×12	9×7	5×12
9×6	4×6	8×4	9×4	5×10	3×7
5×4	7×5	4×9	5×7	7×6	6×10

7. Repeat the table of 6's.
8. Count by 5's from 5 to 60.
9. Repeat the table of 7's from 7 to 84.
10. Count by 3's from 3 to 36.
11. Count by 10's from 10 to 120.
12. Count by 4's from 4 to 48.
13. Repeat the table of 8's; of 9's.

Multiply:

14.	15.	16.	17.	18.
9×10	6×11	7×6	8×4	7×10
8×8	5×12	7×7	9×11	8×7
7×8	7×9	6×12	9×8	9×5
4×9	9×9	5×9	7×11	8×6

Terms in Multiplication. The **sign of multiplication** is $\times$, which is read *times*, or *multiplied by*.

We call the number to be multiplied the **multiplicand.**

The number by which we multiply is called the **multiplier.**

The answer in multiplication is called the **product.**

Multiplication is only a short method of adding numbers. For example, it is quicker to say $5\times 9¢=45¢$ than to say $9¢+9¢+9¢+9¢+9¢=45¢$.

Multiply $246×43.

	CHECK
$246	43
43	246
738	258
984	172
$10,578	86
	$10,578

There are two ways of testing the correctness of our work in multiplication: first, we may go over the work again without rewriting; or second, we may rewrite, putting the multiplicand below, and then multiply again.

WRITTEN EXERCISE

Find the products and prove:

1.	2.	3.	4.	5.	6.
328	453	$379	294	565	$613
21	36	28	32	52	48

7.	8.	9.	10.	11.	12.
3463	392	$4115	813	4216	939
8	36	9	80	7	26

13.	14.	15.	16.
$645×34	832×54	339×45	3049×8
$734×47	936×28	725×62	6170×8

Multiplying Dollars and Cents. How much must be paid for 9 chairs at $12.25 each?

$12.25 9 $110.25	Here we multiply as with whole numbers. We place the decimal point in the product two places from the right under the decimal point in the multiplicand.

WRITTEN EXERCISE

Multiply and prove:

1.	2.	3.	4.	5.
$8.63	$14.39	$25.31	$116.24	$88.76
4	12	17	6	7

6.	7.	8.	9.	10.
$1.35	$22.50	$38.41	$25.06	$94.95
56	47	14	23	18

11. A man's wages are $22.50 a week. Find how much he will earn in 12 weeks.

12. Our rent is $44.25 a month. Find what it will amount to in a year.

13. A dealer sold 33 rugs worth $75 each. Find the amount of the bill.

14. A laborer worked for 16 days at $3.85 a day. How much money did he receive?

Multiplying by 11.

1. Count by 11 from 11 to 132.

11¢+11¢=? 11¢+11¢+11¢=? How much is 6×11¢?

2. How much is 9×11¢? 10×11¢? 11×9¢? 11×10¢?

The Table of 11's. Learn this:

11×1=11	11× 7= 77	1×11=11	7×11= 77
11×2=22	11× 8= 88	2×11=22	8×11= 88
11×3=33	11× 9= 99	3×11=33	9×11= 99
11×4=44	11×10=110	4×11=44	10×11=110
11×5=55	11×11=121	5×11=55	11×11=121
11×6=66	11×12=132	6×11=66	12×11=132

ORAL EXERCISE

1.	2.	3.	4.	5.
7×11	9×11	12×11	4×11	11×10
10×11	6×11	8×11	11×11	11×12

Short Methods of Multiplying by 11.

1. I bought 11 chairs at $28 each. Find the amount of my bill.

\$28 28 \$308	By writing the multiplicand twice, you see that it is unnecessary to write 28 11

2. Try this with 11× {34, 25, 56, 68} 11× {64, 75, 86, 98}

3. Multiply 23 by 11:

23 23 253	This is another short method. We may write 2 3 and then the sum of 2 and 3 in the middle.

4. Try this with 11× { 12, 22, 34, 62, 81 } 11× { 36, 44, 72, 53, 41 }

WRITTEN EXERCISE

Multiply by either short method:

1.	2.	3.	4.	5.
24×11	55×11	94×11	51×11	47×11
35×11	72×11	39×11	93×11	76×11
44×11	84×11	46×11	27×11	82×11

Multiplying by 12.

1. Count by 12's to 144.
2. Give results quickly:

5×12	9×12	12×12	4×12	11×12
7×12	10×12	6×12	3×12	8×12

The Table of 12's. Learn this:

1×12=12	7×12= 84	12×1=12	12× 7= 84
2×12=24	8×12= 96	12×2=24	12× 8= 96
3×12=36	9×12=108	12×3=36	12× 9=108
4×12=48	10×12=120	12×4=48	12×10=120
5×12=60	11×12=132	12×5=60	12×11=132
6×12=72	12×12=144	12×6=72	12×12=144

ORAL EXERCISE

1. How many oranges in 6 dozen? 4 doz.? 1 doz.? 10 doz.? 5 doz.? 11 doz.? 3 doz.? 14 doz.? 8 doz.? 2 doz.? 7 doz.? 9 doz.?

2. How many feet in 2 yards? 4 yd.? 10 yd.? 12 yd.? 5 yd.? 11 yd.? 3 yd.? 8 yd.? 7 yd.? 1 yd.? 9 yd.? 6 yd.?

3. How many quarts in 1 peck? 6 pk.? 2 pk.? 7 pk.? 10 pk.? 4 pk.? 5 pk.? 8 pk.? 9 pk.? 11 pk.? 3 pk.? 12 pk.?

Multiply each number by the multiplier at the left:

4.	2×	12	16	17	19	14	22	18	23	15	21
5.	3×	23	13	22	20	11	16	15	14	12	21
6.	4×	11	12	22	13	21	14	25	15	30	20
7.	3×	15	22	25	21	30	40	11	33	13	31
8.	2×	16	17	24	19	14	25	23	13	26	18

Multiplying Whole Numbers by Multiples of 10 and 100.

1. How much is 10×4? 10×40?
2. Give the product of 10×13. 10×56.
3. Multiply each number by 10:

11	37	62	75	84	95	61
23	45	59	68	91	88	38

$122 \times 10 = 1220$ $\quad$ $82 \times 100 = 8200$

To multiply by 10, *add one zero to the right of the number.*

To multiply by 100, *add two zeros to the right of the number.*

WRITTEN EXERCISE

Multiply by 10; *write the product only:*

1. 237	739	$501	800	1560	$1832
2. 112	631	$456	120	1784	1990

Multiply by 100:

3. 8	12	20	31	45	19	26	52

Multiply by 100, *writing answers only:*

4. 125	$118	275	$300	465	$196
5. 220	$651	311	410	378	204
6. 151	214	383	488	247	182
7. 463	198	722	$500	396	175

Multiply:

8. 443×200	10. $313×200	12. $122×400
9. 232×300	11. 622×300	13. $234×200

Multipliers of Three Figures. Multiply 224 by 156:

224 156 ——— 1344 1120 224 ——— 34944	We write the first partial product with the right-hand figure under the multiplier 6. We write the second partial product with the right-hand figure under the multiplier 5. We write the third partial product with the right-hand figure under the multiplier 1.	PROOF 156 224 ——— 624 312 312 ——— 34944

WRITTEN EXERCISE

Find these products:

1.	2.	3.	4.	5.	6.
222	$334	376	$475	238	217
122	122	221	215	134	116

7.	8.	9.	10.	11.	12.
129	$250	412	325	248	$370
218	226	113	224	123	182

WRITTEN EXERCISE

1. Multiply 227 by 140:

```
  227      We bring down the 0.
  140      Again we write the right-hand figure of the
 ----   partial product under the multiplier 4.
 9080      We place the right-hand figure of the next
227     partial product under the multiplier 1.
-----      Proof: 140×227=31780.
31780
```

2. Multiply 205 by 106:

```
  205      Again we place the right-hand figure of the
  106   partial product under the multiplier 6.
 ----      0×0=0. We place an 0 under the 0.
 1230      We place the right-hand figure of the next
2050    partial product under the multiplier 1.
-----      Proof: 106×205=21730.
21730
```

Find the products:

3.	4.	5.	6.	7.
232	$412	315	128	319
150	120	130	240	220

8. 202 305	9. $307 408	10. 406 209	11. 505 204	12. 409 207
13. 537 103	14. $422 156	15. 262 243	16. 305 209	17. $3.68 2.12
18. 618 209	19. $5.24 104	20. 348 225	21. 484 125	22. $3.18 129

WRITTEN EXERCISE

Multiply:

1. 163×204
2. $4.57×106
3. 118×484
4. 245×154
5. 123×331
6. 234×176
7. 307×308
8. 216×154
9. 312×126
10. 409×105
11. 662×88
12. $1.37×137
13. 276×215
14. $3.54×135
15. 407×248
16. $34.24×60
17. 157×275
18. $1.62×200
19. 206×300
20. 584×87
21. $14.37×40
22. $265×200
23. 145×200
24. $4.69×103
25. 346×129
26. 261×98
27. 536×83
28. 125×154
29. 179×162
30. $21.23×77
31. 245×132
32. 341×118
33. 844×39
34. $8.09×63
35. 638×80
36. 567×30
37. $3.23×122
38. $5.96×64
39. 302×403
40. 1307×48
41. 234×112
42. 345×207
43. 238×216
44. 548×120
45. 924×46
46. 157×145
47. 767×149
48. 182×165
49. 283×256
50. $3.31×309
51. 307×124

ORAL DRILL EXERCISE

	A	*B*	*C*	*D*
1.	$\frac{1}{2}$ of 364	$\frac{1}{5}$ of 505	$\frac{1}{6}$ of 720	$\frac{1}{9}$ of 279
2.	$\frac{2}{3}$ of 330	$\frac{2}{5}$ of 350	$\frac{1}{10}$ of 800	$\frac{2}{3}$ of 360
3.	$\frac{3}{4}$ of 480	$\frac{1}{2}$ of 660	$\frac{2}{3}$ of 420	$\frac{1}{2}$ of 900
4.	54 20	27 30	13 40	28 50
5.	9)20	12)30	13)40	19)40
6.	17)38	16)35	18)40	15)65
7.	7×110	8×120	9×110	11×90
8.	7×210	8×210	9×410	11×70
9.	12)40	13)55	17)70	15)70
Multiply:				
10.	8×8	7×9	8×6	9×9
11.	11×12	84×10	12×12	10×56
12.	7×12	12×30	11×9	11×40
13.	11×10	7×11	12×4	4×11
14.	42×10	30×11	20×12	50×12
15.	99×10	121×10	44×20	132×40
16.	48×10	96×10	144×10	126×30
Add:				
17.	63+23	54+34	72+56	81+72
18.	70+19	14+88	20+31	83+73
19.	53+21	62+45	81+45	48+63
20.	25+27	19+90	49+54	57+21
21.	18+36	93+42	56+26	41+14

GENERAL ORAL PROBLEMS

1. Ice cream is 40¢ a quart. What will 3 pints cost?

2. If 3 cakes cost 10¢, what will a dozen cakes cost?

3. If 2 trunks cost $15, what will half a dozen trunks cost?

4. Harry bought 8 two-cent stamps, 1 five-cent stamp, and 8 postal cards. How much money did he spend?

5. How much change did he get from half a dollar?

6. It takes 3 yards of ribbon to make a bow. What will the ribbon for 2 bows cost at 9¢ a yard?

7. Pens are $.24 a box. Find the cost of $2\frac{1}{2}$ boxes.

8. I paid for the pens with a dollar bill. How much change did I receive?

9. If 2 collars cost a quarter, what will a dozen collars cost?

10. If 1 yard of goods cost $.16, what will 3 yards cost?

11. A girl bought an apron for $.59 and gave the clerk three quarters. How much change did she receive?

12. A storekeeper paid $42 for 7 electric lamps. He sold them for $8.45 each. How much did he gain on each lamp?

13. A newsboy starting out with 64 papers, sold all but $\frac{1}{8}$ of them. How many did he sell?

14. If crackers cost $.18 a box, how much must I pay for 6 boxes?

15. The fruitman buys pineapples at $.96 a dozen. He sells them at $.15 each. How much does he gain on a dozen?

16. Find the cost of 2 yd. of silk if one-half yard cost 45¢.

17. If eggs are 60¢ a dozen, how much must I pay for 48 eggs?

18. A large box containing fruit weighs 23 pounds. If the box alone weighs 4½ pounds, how many pounds does the fruit weigh?

19. If crackers are 25¢ a box, how many boxes can be bought for $4?

20. A man pays $70 a month rent for his store. How much rent does he pay every year?

GENERAL WRITTEN PROBLEMS

1. A bag of coffee weighs 145 pounds. Find the total weight of a truck loaded with 115 bags if the truck weighs 4400 pounds.

2. If each acre of a wheat farm produces 31 bushels, how much wheat will the farmer get from 56 acres?

3. A coal dealer buys coal at $3.35 a ton, and sells it at $4.60 a ton. How much does he make on 15 T.?

4. My mother is buying lace for 3 waists. It costs $.85 a yard and each waist requires 14 yards. Find the cost of the lace.

5. A dealer buys spools of cotton at 36¢ a doz. If he sells it at 4¢ a spool, how much will he gain on 11 dozen spools?

6. A man who earns $2.75 a day, spends $1.30 a day. How much will he save in 7 weeks, if he works 6 days a week?

7. Our rent is $36.25 a month. How much do we pay yearly?

8. A conductor collected 43 fares on his downtown trip and 186 fares on his uptown trip. If each fare is 5¢, how much did he collect on both trips?

9. At 3¢ a quart, find the value of 77 gallons of oil.

10. A dealer had 423 pints of oil. He sold 30 gallons of it. How many gallons has he still to sell?

11. If I pay $5.04 for 9 lb. of tea, how much must I pay for 14 lb. at the same rate?

12. A man who owes $400 has saved $177.28. How much more must he save to pay the debt?

13. A butcher sold to a restaurant 188 lb. beef at $.23 a lb. How much did he receive?

14. A department store ordered $756 worth of willow chairs worth $9 each. How many chairs were received?

15. A tank contained 345 gal. gasoline. 111 gal. were added, and $\frac{3}{4}$ of the contents were sold. How many gal. are left?

16. A workman is paid 28¢ an hour. How much will he receive if he works 49 hours in a week?

VI. DIVISION

Factors. The **factors** of a number are the numbers *which when multiplied together* will produce that number.

$7 \times 5 = 35$. 7 and 5 are factors of 35.

$5 \times 11 = 55$. 5 and 11 are factors of 55.

1. One factor of these numbers is 3; give the other factor:

9	33	27	15	24	36	45	60	90

2. Tell the factors of:

6	18	14	42	45	33	48	24	70
44	10	12	22	28	36	38	46	64
8	50	49	30	25	77	26	90	32
21	16	40	63	80	56	42	54	35

Terms of Division. In division we call the number to be divided the **dividend.**

The number by which we divide is the **divisor.**

The answer in division is called the **quotient.**

When the division is not even or exact, we have a part of the dividend left over. This is the **remainder.**

5)34 *dividend* 6 *quotient,* 4 *remainder.*	Sometimes we write the result as $6\frac{4}{5}$.

We may show division in several ways:

35 divided by 7; $35 \div 7$; $7\overline{)35}$; $7\underline{)35}$.

ORAL EXERCISE

Give the quotients:

1.	4)36	5)30	5)45	6)42	6)66	7)28	7)42
2.	27÷9	44÷11	36÷12	55÷5	72÷8	35÷7	48÷6
3.	9)54	2)48	7)63	11)66	12)72	12)48	3)24

Find the quotients and the remainders:

4.	5)12	3)10	6)14	7)17	10)22	7)29	12)25
5.	4)10	7)15	8)17	9)19	11)24	8)19	9)20
6.	5)27	6)32	7)36	8)26	9)21	8)34	10)32
7.	6)44	8)59	5)56	4)23	7)48	11)57	9)62

Writing a Remainder. Our class uses 5 pieces of chalk every week. If we have 111 pieces in our closet, how many weeks will they last?

5)111 $22\frac{1}{5}$ weeks	Here we have 1 as a remainder. Our chalk will last 22 weeks and 1 over. We write this remainder over the divisor 5, as $\frac{1}{5}$.

WRITTEN EXERCISE

Find the quotients and write the remainders as above:

1. 133÷3	**5.** 632÷5	**9.** 258÷3
2. 473÷4	**6.** 413÷3	**10.** 641÷5
3. 581÷5	**7.** 251÷4	**11.** 583÷6
4. 311÷6	**8.** 273÷6	**12.** 472÷5

Find the quotients and write the remainders as above:

13. 3741÷4	**17.** 2835÷8	**21.** 5473÷7
14. 2627÷5	**18.** 6343÷4	**22.** 2379÷6
15. 3343÷6	**19.** 9763÷3	**23.** 3286÷5
16. 4516÷7	**20.** 8527÷7	**24.** 9263÷4

ORAL EXERCISE

Divide each number by the divisor at the left. State quotients and remainders:

1. 3)	15	29	40	34	25	32	22	20	35	37
2. 2)	15	9	13	21	17	25	35	45	49	19
3. 4)	15	21	33	37	25	22	34	29	50	45

Divide:

4. 6)	13	29	38	19	44	20	33	49	40	27
5. 7)	16	20	36	40	19	48	38	23	44	30
6. 9)	20	30	40	22	29	38	19	47	42	50

Dividing by Multiples of 10.

1. How many 2's in 4? $4 \div 2 = ?$
2. How many 20's in 40? $40 \div 20 = ?$
3. How many 200's in 400? $400 \div 200 = ?$

You see that we get the same answer by canceling or striking out the 0's.

That is: $4 \div 2 = 2.$

$4\not{0} \div 2\not{0} = 2.$

$4\not{0}\not{0} \div 2\not{0}\not{0} = 2.$

WRITTEN EXERCISE

1. Divide 5240 by 40:

$4\not{0})\overline{524\not{0}}$ 131 *quotient.*	When a number ends in 0, we may divide by 10, 20, 30, etc., by marking off or canceling the zero in the dividend and in the divisor.

Divide:

2. 30)360

3. 1540÷70

4. 2320÷80

5. 50)650

6. 70)840

7. 450÷90

8. 1530÷30

9. 40)560

10. 20)1820

11. 2500÷50

12. 1240÷40

13. 30)3390

14. 40)5640

15. 1980÷90

16. 1360÷80

Dividing by Multiples of 10 with Remainders.
Divide 5260 by 40:

$4\not{0})\overline{526\not{0}}$ $131\frac{20}{40}$	We again mark off the 0 in the divisor and in the dividend. We have a remainder of 2 tens or 20 units. Write the complete remainder as 20 units over the complete divisor.

WRITTEN EXERCISE

Divide:

1. 20)4310

2. 7940÷30

3. 6560÷50

4. 40)7830

5. 5230÷30

6. 8500÷70

7. 50)6120

8. 96÷40

9. 3100÷90

10. 60)8530

11. 48÷80

12. 6700÷60

WRITTEN EXERCISE

1. Divide 5241 by 40:

$4\not{0}\overline{)524\not{1}}$	We again mark off the 0 in the divisor and the units' figure in the dividend.
$4\not{0}\overline{)524\not{1}}$ $\frac{1}{40}$	Write this remainder, the units' figure, over the complete divisor 40.
$4\not{0}\overline{)524\not{1}}$ $131\frac{1}{40}$	Divide 524 by 4.

Divide:

2. $30\overline{)2162}$
3. $40\overline{)4843}$
4. $5054 \div 50$
5. $1476 \div 70$
6. $20\overline{)8465}$
7. $50\overline{)1052}$
8. $4979 \div 70$
9. $8088 \div 80$
10. $30\overline{)9277}$
11. $40\overline{)8286}$
12. $3555 \div 50$
13. $2874 \div 70$
14. $60\overline{)1863}$
15. $60\overline{)2461}$
16. $5496 \div 90$

WRITTEN EXERCISE

1. Divide 5261 by 40:

$4\not{0}\overline{)526\not{1}}$ $\frac{1}{40}$	Here again we mark off the 0 in the divisor and the units' place in the dividend. Write the units' figure 1 over the complete divisor 40.
$4\not{0}\overline{)526\not{1}}$ $131\frac{21}{40}$	Divide by 4, and write the remainder 2 tens before the 1, making the complete remainder 21.

Divide:

2. 20)3707	6. 90)5804	10. 30)2929
3. 40)5308	7. 80)6303	11. 70)1851
4. 60)9705	8. 50)7478	12. 50)6325
5. 70)4303	9. 30)4757	13. 40)9104

Division Tables of 11's and 12's.

1. Tell the answers:

$\frac{1}{11}$ of 22	$\frac{1}{11}$ of 55	$\frac{1}{12}$ of 24	$\frac{1}{11}$ of 66	$\frac{1}{12}$ of 60
$\frac{1}{11}$ of 77	$\frac{1}{12}$ of 36	$\frac{1}{12}$ of 48	$\frac{1}{12}$ of 72	$\frac{1}{12}$ of 84
$\frac{1}{11}$ of 11	$\frac{1}{12}$ of 12	$\frac{1}{11}$ of 33	$\frac{1}{11}$ of 88	$\frac{1}{11}$ of 44
$\frac{1}{11}$ of 110	$\frac{1}{12}$ of 120	$\frac{1}{11}$ of 121	$\frac{1}{12}$ of 96	$\frac{1}{11}$ of 99

2. Learn these tables:

11 ÷ 11 = 1	77 ÷ 11 = 7	12 ÷ 12 = 1	84 ÷ 12 = 7
22 ÷ 11 = 2	88 ÷ 11 = 8	24 ÷ 12 = 2	96 ÷ 12 = 8
33 ÷ 11 = 3	99 ÷ 11 = 9	36 ÷ 12 = 3	108 ÷ 12 = 9
44 ÷ 11 = 4	110 ÷ 11 = 10	48 ÷ 12 = 4	120 ÷ 12 = 10
55 ÷ 11 = 5	121 ÷ 11 = 11	60 ÷ 12 = 5	132 ÷ 12 = 11
66 ÷ 11 = 6	132 ÷ 11 = 12	72 ÷ 12 = 6	144 ÷ 12 = 12

3. State the quotients rapidly:

99÷11	110÷11	121÷11	132÷11	88÷11
72÷12	44÷11	96÷12	108÷12	22÷12
84÷12	48÷12	144÷12	60÷12	120÷12
55÷11	36÷12	33÷11	66÷11	12÷12
24÷12	108÷12	11÷11	132÷12	77÷11

Steps in Long Division.

1. Divide 192 by 2:

2)192 *dividend* 96 *quotient*	This way of working division is called **Short Division.**

2. Divide 192 by 12:

16 *quotient* 12)192 12 72 72	This longer way of dividing in which we write down all the remainders is called **Long Division.**

3. Divide 682 by 22:

31 *quotient* 22)682 66 22 22	(1) To get the first quotient figure we try dividing 6 by 2. This gives 3 as the number of times 22 is contained in 68 the partial dividend. (2) Write this 3 in the quotient above the 8 of 68. (3) $22 \times 3 = 66$. Write this under 68. (4) $68 - 66 = 2$. (5) Bring down the next figure of the dividend, 2.

$22 \div 22 = 1$ Write the 1 in the quotient, over the 2.

$22 \times 1 = 22$ Subtract $22 - 22 = 0$.

Proof: If our work has been correct, the quotient multiplied by the divisor should give the dividend.

$$31 \times 22 = 682.$$

This gives us the five steps in Long Division.

(1) We *find the quotient figure by dividing.*

(2) We *write this figure in the quotient.*

(3) We *multiply the divisor by it.*

(4) We *subtract the product from the partial dividend.*

(5) We *bring down the next figure.*

WRITTEN EXERCISE

Divide in this way and prove:

1. 11)121	**8.** 21)441	**15.** 22)682	**22.** 32)992
2. 12)144	**9.** 23)483	**16.** 31)961	**23.** 33)693
3. 21)651	**10.** 11)191	**17.** 44)924	**24.** 21)7518
4. 21)441	**11.** 31)651	**18.** 35)735	**25.** 26)3886
5. 21)882	**12.** 33)768	**19.** 26)546	**26.** 41)7626
6. 22)681	**13.** 21)897	**20.** 37)777	**27.** 23)3776
7. 24)288	**14.** 24)528	**21.** 28)588	**28.** 32)4256

How to Divide Dollars and Cents.

Divide $45.64 by 7:

7)$45.64
$6.52

When we are dividing numbers representing dollars and cents, we place the decimal point in the quotient exactly below or above the decimal point in the dividend.

```
   $6.52
7)$45.64
   42
    36
    35
     14
     14
```

WRITTEN EXERCISE

Divide:

1. $\$97.35 \div 3$
2. $\$87.44 \div 4$
3. $\$79.32 \div 2$
4. $\$58.14 \div 9$
5. $\$36.54 \div 3$
6. $\$39.84 \div 8$
7. $\$53.83 \div 7$
8. $\$96.10 \div 5$
9. $\$411.06 \div 6$
10. $\$28.45 \div 5$
11. $\$411.76 \div 4$
12. $\$713.28 \div 4$
13. $\$28.35 \div 9$
14. $\$91.60 \div 8$
15. $\$824.31 \div 9$

Long Division Continued.

Divide 8526 by 42:

```
    203
42)8526
   84
   ---
    126
    126
    ---
```

The first quotient is 2. We multiply and subtract and have 1 as a remainder.

Here we bring down the 2. We have 12 to be divided by 42.

But 12 does not contain 42, so we place 0 in the quotient above the 2. On bringing down the 6, we have a partial dividend that contains 42.

WRITTEN EXERCISE

Divide these in the same way and prove:

1. $1224 \div 12$
2. $3132 \div 29$
3. $2354 \div 22$
4. $6420 \div 30$
5. $4598 \div 22$
6. $4968 \div 24$
7. $2320 \div 40$
8. $7337 \div 23$
9. $9840 \div 41$
10. $7200 \div 50$
11. $4033 \div 37$
12. $9386 \div 26$
13. $3900 \div 60$
14. $8000 \div 32$
15. $8330 \div 70$
16. $4387 \div 41$
17. $3706 \div 34$
18. $5992 \div 56$
19. $1962 \div 18$
20. $8772 \div 43$
21. $4212 \div 39$
22. $2332 \div 22$
23. $6624 \div 32$
24. $8056 \div 76$

WRITTEN EXERCISE

1. Divide 6888 by 22:

$313\frac{2}{22}$ *quotient* 22)6888 66 28 22 68 66 2 *remainder*	We divide in the usual way. We find that 22 is contained 3 times in 68, with a remainder of 2. Notice where this is written.	PROOF 313 22 626 626 6886+2=6888.

Find these quotients:

2. 735÷32
3. 332÷14
4. 936÷41
5. 762÷52
6. 391÷18
7. 897÷22
8. 4334÷31
9. 2999÷27
10. 6987÷32
11. 4798÷34
12. 7358÷35
13. 4798÷34
14. 7656÷25
15. 7920÷25
16. 9883÷24

WRITTEN EXERCISE

1. Divide 1240 by 22.

$56\frac{8}{22}$ 22)1240 110 140 132 8	Here we find we cannot divide 12 by 22. In this case we use the first three figures, 124, as the partial dividend.

Divide:

2. 1512÷21	**7.** 1324÷42	**12.** 1085÷51
3. 1323÷21	**8.** 1475÷62	**13.** 2224÷32
4. 1546÷25	**9.** 1122÷25	**14.** 1518÷37
5. 518÷24	**10.** 1864÷45	**15.** 1345÷42
6. 2193÷51	**11.** 1996÷32	**16.** 1982÷31

ORAL EXERCISE

1. Divide rapidly:

4)280	5)550	7)350	12)600	5)600	9)270
9)810	8)640	7)490	6)420	8)400	9)720
12)480	11)880	8)560	11)330	7)630	6)600

2. How many weeks in 21 days? In 49 days? In 63 days? In 14 days? In 28 days? In 70 days? In 84 days? In 77 days? In 35 days? In 42 days? In 56 days? In 140 days?

3. How many pecks in 24 quarts? In 40 quarts? In 80 quarts? In 88 quarts? In 56 quarts? In 64 quarts? In 96 quarts? In 16 quarts? In 72 quarts? In 32 quarts? In 48 quarts?

4. Divide each of the following numbers by the left-hand figure; give quotients and remainders:

11)24	56	67	100	36	79	30	45	27	80
6)19	66	34	59	74	25	20	41	36	55
9)29	65	20	75	30	47	40	93	38	100

5. At 5¢ each, tell how many carfares I can pay with:

60¢	35¢	55¢	75¢	90¢	$1.00	65¢
25¢	$1.15	40¢	85¢	$1.25	$2.00	$1.45

6. Divide:

12)37	10)53	12)66	10)95	12)50	12)75
20)105	60)124	30)122	50)506	40)202	60)185
10)83	11)70	12)30	10)48	11)100	12)27

Trial Divisors.

1. Divide 9161 by 29:

$$\begin{array}{r} 315\frac{26}{29} \\ 29\overline{)9161} \\ \underline{87} \\ 46 \\ \underline{29} \\ 171 \\ \underline{145} \\ 26 \end{array}$$

In this case it is not very easy to find the first figure of the quotient. 29, however, is very near 30; so we use 30 as a *trial divisor.*

$91 \div 30 = 3.$

2. Tell what number you would use as a trial divisor, in order to find the first figure of the quotient in these cases:

91)	72)	52)	31)	12)	61)	17)
59)	19)	28)	77)	88)	27)	96)

WRITTEN EXERCISE

Divide:

1. 910÷19	**5.** 2136÷48	**9.** 412÷79
2. 6302÷28	**6.** 9013÷27	**10.** 8452÷38
3. 9912÷48	**7.** 1274÷57	**11.** 2465÷59
4. 6394÷29	**8.** 8375÷77	**12.** 2165÷49

13. Divide 1734 by 23:

$$\begin{array}{r} 75\frac{9}{23} \\ 23\overline{)1734} \\ 161 \\ \hline 124 \\ 115 \\ \hline 9 \end{array}$$

Here if we try dividing 17 by 2 or 173 by 20, we have 8 as the first quotient figure.

But $8 \times 23 = 184$, which is greater than the partial dividend 173, so we try 7. Again if we try dividing 12 by 2 or 124 by 20, we have 6. $6 \times 23 = 138$, a product which is greater than the partial dividend 124.

No product can be larger than the partial dividend.

14. Divide 1394 by 27:

$$\begin{array}{r} 51\frac{17}{27} \\ 27\overline{)1394} \\ 135 \\ \hline 44 \\ 27 \\ \hline 17 \end{array}$$

In this case since 27 is near 30, we might say $139 \div 30 = 4$.

$4 \times 27 = 108$, and $139 - 108 = 31$.

But 31 is larger than 27, and this means that 4 is too small. We must use 5.

No remainder can be larger than the divisor.

WRITTEN EXERCISE

Divide:

1. 9147÷48	**7.** 7344÷13	**13.** 8156÷76
2. 6524÷41	**8.** 2763÷80	**14.** 2266÷38
3. 7785÷30	**9.** 5475÷77	**15.** 1449÷50
4. 8152÷19	**10.** 1837÷23	**16.** 1673÷24
5. 832÷26	**11.** 3165÷21	**17.** 2476÷16
6. 5040÷72	**12.** 6946÷30	**18.** 3796÷26

Divide and prove:

19. 4768÷62	**23.** 9464÷81	**27.** 3259÷67
20. 3453÷57	**24.** 6040÷38	**28.** 5863÷74
21. 9934÷45	**25.** 7659÷42	**29.** 6294÷85
22. 7863÷39	**26.** 8136÷56	**30.** 2356÷93

WRITTEN EXERCISE

Divide:

1. 7970÷33	**13.** 6456÷26	**25.** 8920÷15
2. 3792÷21	**14.** 5250÷35	**26.** 3348÷11
3. 3456÷36	**15.** 5508÷32	**27.** 5830÷30
4. 2948÷23	**16.** 2330÷24	**28.** 2860÷25
5. 2664÷18	**17.** 2944÷23	**29.** 5632÷40
6. 2842÷29	**18.** 4992÷39	**30.** 4098÷32
7. 4260÷28	**19.** 5432÷28	**31.** 8920÷80
8. 4189÷59	**20.** 3840÷59	**32.** 3192÷28
9. 2378÷29	**21.** 2088÷24	**33.** 8328÷73
10. 5725÷59	**22.** 3498÷23	**34.** 5952÷32
11. 2970÷28	**23.** 6132÷84	**35.** 3469÷62
12. 4802÷49	**24.** 2838÷36	**36.** 2638÷47

WRITTEN EXERCISE

Divide:

1. 9936÷138	11. 31,644÷131	21. 264,945÷342
2. 3469÷122	12. 22,112÷103	22. 869,036÷421
3. 5964÷225	13. 36,737÷225	23. 383,259÷218
4. 8376÷349	14. 61,280÷141	24. 450,887÷322
5. 3144÷131	15. 82,184÷330	25. 246,347÷212
6. 9650÷386	16. 25,833÷123	26. 447,055÷233
7. 46,420÷222	17. 35,850÷107	27. 314,675÷329
8. 159,020÷343	18. 76,874÷133	28. 358,362÷2380
9. 409,285÷129	19. 158,976÷207	29. 844,860÷120
10. 621,376÷112	20. 914,332÷392	30. 922,186÷1140

WRITTEN REVIEW PROBLEMS

1. An army agent is given $7310 to purchase mules. How many can he buy at $215 each?

2. A boat can sail 7 miles an hour. How long will it take to sail 1344 miles at this rate?

3. How many 1-pound packages of tea can a grocer make up from 608 oz.?

4. The army bought 32 horses for $5440. Find the price paid for the average horse.

5. The janitor received $1872 for taking care of the school building last year. How much did he receive each month?

6. How many bushels are there in 336 pecks?

7. The divisor is 38, the dividend 398. Find the quotient.

8. If 9 trees are planted in each row, how many rows will 2574 trees make?

9. How many dozen oranges can a dealer get from a load containing 28,476 oranges?

10. 8232 pounds of sugar were placed in 21 barrels. How many pounds were put in each barrel?

11. If 32 armchairs are worth $864, what will be the cost of 4 doz. chairs at the same price?

12. A dealer sold 114 gal. of olive oil at 38¢ a qt. How much did he receive for it?

13. He sold to one store 78 one-gallon cans for $234, and to another store $468 worth of oil at the same rate. How many cans were shipped to the second store?

14. A manufacturer made 42 doz. coats and sold $\frac{1}{4}$ of them at $7 each, wholesale. How much did he receive for the coats he sold?

15. If $464 is paid for 29 bicycles, how many can be bought for $240?

16. A girl found that in 47 days she had worked 1081 problems in arithmetic. Find the average number she worked per day.

17. A storekeeper's income is $2340 a year. How much does his store pay him each week?

18. A furniture dealer paid $2268 for dining-room chairs at $12 each. How many chairs did he buy?

19. If a bu. of oats weighs 32 lb., how many bu. are there in a load of oats weighing 14,400 lb.?

ORAL DRILL EXERCISE

	A	B	C	D
1.	7)840	8)720	9)810	9)540
2.	7)560	11)550	9)360	12)720
3.	10)900	12)480	11)330	8)320
4.	600÷5	900÷3	1200÷4	1500÷3
5.	7)490	12)960	8)640	9)630

Give results:

	A	B	C	D
6.	$\frac{1}{2}$ of 48	11×12	11×?=110	11×11
7.	12)132	$\frac{1}{3}$ of 72	?)108 12	?)110 11
8.	17×?=34	?)121 11	18×?=36	9×?=108
9.	$\frac{1}{12}$ of 144=?	12×?=84	$\frac{1}{4}$ of 480	$\frac{1}{10}$ of 700
10.	10×?=120	10×12	11×?=66	12×12

Add:

	A	B	C	D
11.	72+12	81+19	90+27	54+36
12.	15+22	48+84	26+59	64+73
13.	91+93	35+33	44+56	56+87
14.	23+87	14+56	19+17	42+32
15.	14+62	54+19	45+26	32+16

Subtract:

	A	B	C	D
16.	$14.22 80	$16.37 40	$8.45 90	$19.21 70
17.	$27.36 80	$54.63 70	$72.81 90	$94.83 90

GENERAL ORAL PROBLEMS

1. How much would you pay for a toothbrush at 25¢, and 3 pieces of soap at 10¢ each?

2. A string is 72 inches long. What is its length in feet?

3. Harry bought a pair of skates for $2.25. How much change did he get from a five-dollar bill?

4. A company marches 72 miles in 8 hours. How many miles an hour do they march?

5. Tell the cost of:

10 things, if 2 cost 5¢
9 things, if 3 cost 8¢
14 things, if 7 cost 35¢
24 things, if 8 cost 6¢
7 things, if 3 cost 12¢
5 things, if 2 cost 14¢

6. The grocer had 19 pecks of potatoes and bought 36 pecks more. How many has he now?

7. George bought 16 2-cent stamps and 8 postal cards. How much change will he get from a half dollar?

8. Ethel has 32 two-cent stamps. How many letters can she mail if each one takes 4¢ in stamps?

9. If the boys use the basket ball 15 minutes a day, how many hours will they use it in 5 days?

10. If they play 3 games each week, how many games will they play in a month and a half?

11. How many gallons of milk in a can containing 36 quarts?

12. At $11 each, how many sofas can be bought for $121?

13. A dealer paid $65.80 for a showcase, and $57.50 for an electric sign. How much more did the case cost than the sign?

14. In printing 660 report cards, $\frac{1}{6}$ were printed the wrong way. How many were done the right way?

15. How many one-ounce bags can a grocer fill from 2 pounds of pepper?

GENERAL WRITTEN PROBLEMS

How to Solve Problems: Cancellation.

1. If 4 lamps cost $5, what will 44 cost at this rate?

> A quick way to write this problem is to use the horizontal line between the dividend and the divisor. This line indicates division.
>
> $$\frac{5\times \overset{11}{\cancel{44}}}{\cancel{4}} = \$55.$$
>
> 1 lamp will cost $\frac{1}{4}$ of $5 or $\frac{5}{4}$.
> 44 lamps will cost $\frac{44\times 5}{4}$
>
> Here we divide 44 by 4, canceling both.

2. If the Field Day expenses of 15 boys are $.12 each, what would the expenses of 40 boys be at the same rate?

3. 7 classes use 11 boxes of chalk in one term. At this rate how many boxes will be needed for 21 classes?

4. A grocer bought 7 barrels of potatoes for $28. How much would he pay for 22 barrels at this rate?

5. 54 army horses were carried in 3 railroad cars. How many horses could be carried in 17 cars?

6. 9 of these horses were worth $1287. Find the value of all 54 horses at the same price.

7. Find how much 21 electric lamps would cost, if $18 was paid for 7.

8. A dealer asked $72 for 9 pictures. The full set contains 15 pictures. What would he ask for a set at this rate?

9. What will 9 overcoats cost, if $333.13 is the cost of 7?

10. The furniture dealer buys chairs at $28.20 a doz. At this rate how much would 11 cost him?

11. Helen paid the laundryman 90¢ for doing 15 pieces. How much would she have to pay for 70 pieces at this rate?

12. If it requires 3 hours' work to finish a pair of curtains, how many hours will be needed to finish 27 curtains?

13. It requires 3 hours' work to finish 2 jackets. How many jackets can be finished in 72 hours?

14. At the rate of 355 miles in 25 hours, how many miles would an automobile go in 15 hours?

15. How many yards of cotton goods at 16 cents a yard will pay a farmer for 34 pounds of butter at 24 cents a pound?

VII. FRACTIONS

Finding Fractional Parts of Numbers.

1. If I give out 28 sheets of paper to 4 boys. how many sheets will each boy receive?

> Here we have to divide 28 into 4 equal parts. We write:
>
> $$28 \div 4, \text{ or } 4\overline{)28}, \text{ or } 4\underline{)28}$$
>
> If we divided the paper into two parts, what would each part be called? When we divide it into 4 parts, each part is called *one fourth* or *one quarter* and it is written $\frac{1}{4}$.

2. Into how many parts should we have to divide the paper in order to have $\frac{1}{3}$, $\frac{1}{5}$, $\frac{1}{6}$, $\frac{1}{7}$, $\frac{1}{8}$, $\frac{1}{10}$?

These expressions which tell into how many equal parts something has been divided are called **fractions.**

3. Divide 36 by 4. What is $\frac{1}{4}$ of 36? Divide 32 by 8. What is $\frac{1}{8}$ of 32?

4. Divide 42 by 6. What is $\frac{1}{6}$ of 42? Divide 50 by 5. What is $\frac{1}{10}$ of 50?

When we say $\frac{1}{4}$ of 32, we wish to find only 1 of the 4 equal parts of 32. When we say $\frac{3}{4}$, we wish to find three of the 4 equal parts of 32. $\frac{7}{10}$ of 40 means 7 of the 10 equal parts of 40; that is, 28.

5. Read these examples and give the results:

$\frac{1}{2}$ of 20	$\frac{1}{4}$ of 12	$\frac{1}{6}$ of 12	$\frac{1}{6}$ of 36	$\frac{1}{10}$ of 10
$\frac{1}{3}$ of 90	$\frac{3}{4}$ of 12	$\frac{1}{8}$ of 24	$\frac{1}{3}$ of 90	$\frac{4}{10}$ of 40
$\frac{2}{3}$ of 9	$\frac{1}{5}$ of 25	$\frac{1}{2}$ of 48	$\frac{2}{3}$ of 90	$\frac{3}{10}$ of 100
$\frac{1}{4}$ of 28	$\frac{2}{5}$ of 25	$\frac{1}{9}$ of 18	$\frac{1}{2}$ of 100	$\frac{8}{10}$ of 100

6. Write in figures: one fourth; two fifths; five sixths; one eighth; four sevenths; seven tenths; five ninths.

ORAL EXERCISE

1. If I wish to find $\frac{1}{5}$ of this line, into how many equal parts must I divide it? To find $\frac{3}{5}$ of the line, how many of these parts must I take?

2. How many dots do you count?

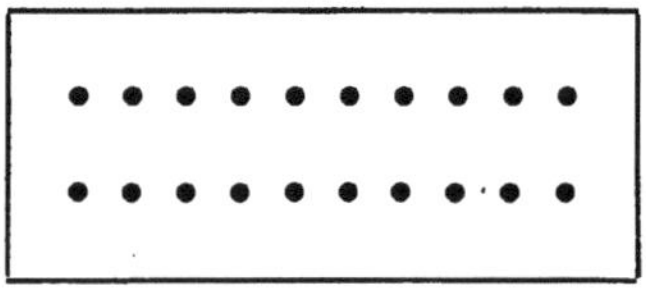

How can I find $\frac{1}{5}$ of these dots?

How many in $\frac{1}{5}$? In $\frac{2}{5}$? In $\frac{3}{5}$? In $\frac{4}{5}$?

How do I find $\frac{1}{5}$ of a thing? $\frac{1}{5}$ of a group of things?

How do I find $\frac{3}{5}$ of a thing? $\frac{3}{5}$ of a group of things?

3. Find $\frac{2}{3}$ of 9; 18; 36; 24; 30; 90; 12; 15.

4. Find $\frac{3}{4}$ of 4; 8; 24; 40; 44; 16; 32; 36.

5. Find $\frac{1}{6}$ of 6; 12; 18; 30; 36; 24; 60; 42.

6. Find $\frac{5}{6}$ of 6; 12; 18; 30; 36; 24; 60; 42.

7. Find $\frac{1}{5}$ of 10; 20; 40; 35; 45; 50; 55; 60.

8. Take the ruler. How many inches are there in $\frac{1}{2}$ of a foot? In $\frac{1}{3}$ foot? In $\frac{2}{3}$ foot? In $\frac{1}{4}$ foot? In $\frac{3}{4}$ foot? In $\frac{1}{6}$ foot? In $\frac{5}{6}$ foot?

How do I find $\frac{1}{6}$ of a foot? What must I do to find $\frac{5}{6}$ of a foot?

9. What is $\frac{1}{8}$ of 16? 32? 24? 48? 56? 80?

How do you find $\frac{3}{8}$ of a number? How would you find $\frac{7}{8}$? $\frac{5}{8}$?

10. Find $\frac{3}{8}$ of each of the numbers in Ex. **9**.

11. What is $\frac{3}{8}$ of 32? 8? 24? 16? 40? 80? 48? 56? 64?

12. What is $\frac{5}{8}$ of \$40? \$15? \$80? \$72? \$32? \$24? \$56? \$8? \$48?

13. Find $\frac{7}{8}$ of 8; 48; 64; 32; 16; 80; 24; 56; 72.

WRITTEN EXERCISE

1. Find $\frac{5}{8}$ of 256:

$\frac{1}{8}$ of 256 = 32	A quicker way is to write:
$\frac{5}{8}$ of 256 = 32 × 5 = 160	8)256 32 × 5 = 160

2. Find $\frac{1}{4}$ of 72; 56; 120; 140; 256; 320; 88; 100.

3. Find $\frac{1}{6}$ of 54; 90; 132; 156; 186; 198; 252; 288.

4. Find $\frac{2}{3}$ of 21¢; 27¢; 33¢; 45¢; 42¢; \$.90; \$1.08; \$1.53; \$1.83.

5. Find $\frac{5}{6}$ of 48; 54; 72; 144; 216; 252; 324; 432.

ORAL EXERCISE

Give answers rapidly:

1. $\frac{2}{3}$ of:	30	24	27	21	33	18	12	36
2. $\frac{3}{4}$ of:	44	16	28	12	48	16	8	32
3. $\frac{7}{8}$ of:	64	32	48	72	88	96	80	56
4. $\frac{3}{5}$ of:	50	35	40	55	5	60	45	20
5. $\frac{5}{6}$ of:	30	12	54	18	72	48	36	60
6. $\frac{3}{7}$ of:	21	49	35	63	56	77	28	42
7. $\frac{3}{8}$ of:	56	8	64	88	32	16	40	24

WRITTEN EXERCISE

1. Find $\frac{2}{3}$ of:	\$54	\$153	\$96	\$105
	\$243	\$174	\$228	\$282
2. Find $\frac{3}{4}$ of:	324 pt.	\$172	376 qt.	304 gal.
	304 pt.	\$204	248 qt.	352 pk.
3. Find $\frac{3}{8}$ of:	72 books	144 men	320 sheep	\$256
	280 lb.	344 oz.	408 doz.	496 hr.
4. Find $\frac{5}{6}$ of:	108 yd.	306 gal.	192 ft.	210 in.
	432 da.	84 wk.	216 mo.	444 yr.

5. Find $\frac{2}{3}$ of:	369	336	414	198	318	249
6. Find $\frac{3}{4}$ of:	288	376	332	456	440	656
7. Find $\frac{3}{5}$ of:	810	585	890	865	345	725
8. Find $\frac{5}{6}$ of:	600	378	300	330	294	372
9. Find $\frac{3}{8}$ of:	264	352	272	496	176	592
10. Find $\frac{7}{8}$ of:	184	576	568	368	504	848
11. Find $\frac{2}{7}$ of:	497	385	224	504	301	448
12. Find $\frac{4}{9}$ of:	396	828	504	756	432	702

ORAL DRILL EXERCISE

	A	*B*	*C*	*D*
State results:				
1.	35¢+85¢	25¢+95¢	65¢+75¢	45¢+35¢
2.	15¢+95¢	75¢+35¢	35¢+85¢	15¢+95¢
3.	70¢−35¢	60¢−25¢	90¢−75¢	80¢−45¢
4.	40¢−15¢	70¢−55¢	80¢−35¢	90¢−55¢
5.	35¢−25¢	80¢−45¢	90¢−15¢	70¢−45¢
6.	$\frac{3}{4}$ of 64	$\frac{1}{3}$ of 69	$\frac{1}{2}$ of 86	$\frac{1}{4}$ of 64
7.	9×210	$\frac{2}{3}$ of 63	6×510	4×820
8.	$\frac{1}{2}$ of 168	7×610	$\frac{1}{2}$ of 250	$\frac{3}{4}$ of 36
9.	$\frac{2}{3}$ of 96	5×910	$\frac{2}{3}$ of 90	2×1220
10.	6×610	$\frac{1}{4}$ of 484	4×310	$\frac{1}{6}$ of 420
Find quotients:				
11.	$10\overline{)580}$	$20\overline{)240}$	$90\overline{)990}$	$50\overline{)350}$
12.	$60\overline{)720}$	$10\overline{)460}$	$40\overline{)480}$	$80\overline{)560}$
13.	$30\overline{)450}$	$30\overline{)650}$	$40\overline{)600}$	$80\overline{)640}$
14.	$20\overline{)340}$	$50\overline{)700}$	$70\overline{)490}$	$10\overline{)490}$
Give answers:				
15.	65÷4	39÷8	75÷6	85÷9
16.	16×3	$\frac{3}{4}$ of 120	44÷?=22	46÷2
17.	$\frac{1}{15}$ of 60	3×?=42	$\frac{1}{18}$ of 36	65÷7
18.	82÷7	$\frac{2}{3}$ of 66	85÷7	$\frac{2}{5}$ of 55
19.	?÷24=2	54÷11	$\frac{1}{7}$ of 560	$\frac{1}{17}$ of 34
20.	20+30+45	70+30+35	40+60+65	60+20+85
21.	60+20+25	20+90+55	20+50+55	30+20+55

GENERAL ORAL PROBLEMS

1. I drew a line 20 inches long and then made it $\frac{3}{5}$ longer. How long did I make the line?

2. Harry paid \$.12 for 1 dozen picture cards. He sold them at 3 for 5¢. How much did he gain?

3. A ceiling is 108 inches high. How many feet in height is it?

4. If 4 bells ring every 3 hours, how many will ring between 9 A.M. and 3 P.M.?

5. Bananas are 3 for 5¢. What will a dozen cost?

6. If they were 4 for 5¢, how much would I save on a dozen?

7. If 5 notebooks cost \$.12, what will 25 cost?

8. What is the difference between $\frac{2}{3}$ of 12 and $\frac{3}{5}$ of 10?

9. If a quart of potatoes costs 9¢, what will a peck cost?

10. If a pound of butter cost 28¢, what will $1\frac{1}{2}$ pounds cost?

11. A farmer bought a cow for \$48 and sold it for $\frac{1}{8}$ more than the cost. How much did he get for it?

12. If $\frac{3}{4}$ of a pound of candy costs 27¢, what will a one-pound box cost?

13. A train takes 3 days to make a trip between two cities. How many hours does it take?

14. Find the cost of 96 apples at 10¢ a dozen.

15. At 24¢ a pound, how much will $2\frac{1}{2}$ pound of crackers cost?

16. A trolley car carried 71 passengers. At 5¢ each, how much did the fares amount to?

17. A man has $480 in a bank. If he should draw out $\frac{5}{6}$ of his money, how much would he leave in the bank?

18. Find the cost of 18 pears at 36¢ a dozen.

GENERAL WRITTEN PROBLEMS

1. $\frac{5}{8}$ of the 328 pupils in a school were boys. How many boys were there in the school?

2. A man placed $85,000 in a bank and later drew out three fifths of it. How much did he leave in the bank?

3. $\frac{5}{6}$ of a shipment of 1200 barrels of flour came to New York. How many barrels were sent to New York?

4. There were 848 men in a regiment. $\frac{7}{8}$ of them were under 40 years of age. How many were under this age?

5. $\frac{4}{5}$ of a shipload of 865 barrels were barrels of cement. How many barrels of cement were there in the shipload?

6. Of a carload of 264 barrels of cranberries, $\frac{3}{8}$ of them were sold at $11.20 a barrel. How much payment was received for this part?

7. A grocer bought $\frac{3}{4}$ of a shipment of 528 boxes of oranges at $4.80 a box. How much did they cost him?

8. There are 624 electric lights in the school. Three eighths of them need new bulbs. How many new bulbs are required?

9. $\frac{6}{10}$ of a load of 120 gallons of olive oil were sold at 48¢ a quart. How much was received for it?

10. Every morning on his way to school, Harry walks $\frac{2}{3}$ of the distance of 369 yards. How many feet does he walk?

11. Mr. Smith had $1584 in the bank. He drew out $\frac{3}{8}$ of it and added $125 to this amount to buy an automobile. How much did it cost?

12. Out of a total of 252 days, $\frac{5}{6}$ were clear days. How many were stormy?

13. A dealer paid $93.80 for 7 willow chairs. Find the cost of each chair.

14. The dealer above sold each chair at a gain of $3.25. How much did he receive for them all?

15. He drew $854.49 from his bank and paid $\frac{3}{7}$ of it for rent. How much is his rent?

16. How much will he receive for 8 dozen rockers at $7.75 each?

17. A restaurant keeper ordered 11 bushels of potatoes at $.28 a peck. Find the amount of his bill.

18. A reel of wire contains 4488 feet. How many yards in this reel?

19. A landlord received in rents $663 a month. If he had 17 apartments in the building paying the same rent, find how much he received from each.

VIII. MEASURES

Length Measure.

1. How many feet in 2 yd.? How many inches in $\frac{1}{2}$ yd.?

2. Measure the width of this book; your desk; the teacher's desk; the blackboard; the room. What units did you use in each case?

3. What street is about one mile north of the school?

Table of Length.

12 inches (in.)	=1 foot (ft.)
3 feet or 36 inches	=1 yard (yd.)
1760 yd. or 5280 feet	=1 mile (mi.)

Reduction of Compound Numbers. Reduce.

1. 4 ft. 8 in. to in.

$$\begin{array}{r} 12 \\ 4 \\ \hline 48 \text{ in.} \\ 8 \\ \hline 56 \text{ in. in 4 ft. 8 in.} \end{array}$$

2. 9 yd. 2 ft. to ft.

$$\begin{array}{r} 3 \\ 9 \\ \hline 27 \text{ ft.} \\ 2 \\ \hline 29 \text{ ft. in 9 yd. 2 ft.} \end{array}$$

WRITTEN EXERCISE

Reduce:

1. 3 ft. 5 in. to inches.
2. 5 ft. 4 in. to inches.
3. 2 yd. 2 ft. to feet.
4. 7 yd. 1 ft. to feet.

5. Change 1 mi. 200 ft. to feet.
6. How many feet in 14 yd.?
7. Change 3 mi. 760 ft. to feet.
8. Change 26 yd. 2 ft. to feet.
9. Change 8 mi. 111 yd. to yards.
10. Change 3 yd. 2 ft. 2 in. to inches.

Reducing to Higher Denominations.

Reduce 86 feet to yards and feet:

3)86
28 yd. 2 ft. in 86 ft.

WRITTEN EXERCISE

Reduce:

1. 49 in. to feet and inches.
2. 137 ft. to yards and feet.
3. 354 in. to yards and feet.
4. 1924 yd. to miles and yards.
5. 4380 ft. to yards and feet.
6. 7254 yards to miles and yards.
7. 1997 feet to yards and feet.

Measures of Weight. How many ounces in 1 pound? In 3 lb.? Meat, candy, and many things in the grocery store are sold by the **pound.** Name some things sold by the **ounce.**

The butcher buys his meat, and the stableman his feed, by the **hundredweight.** Hay, coal, brick, iron and stone are sold by the **ton.**

Table of Weight.

16 ounces (oz.)	=1 pound (lb.)
100 pounds	=1 hundredweight (cwt.)
2000 pounds	=1 short ton (T.)
2240 pounds	=1 long ton (L. T.)

WRITTEN EXERCISE

Reduce:

1. 20 lb. to ounces.
2. 8 cwt. to ounces.
3. 4 cwt. 22 lb. to pounds.
4. 11 lb. 9 oz. to ounces.
5. 4 short tons to ounces.
6. 6 long tons 17 lb. to pounds.
7. 11 cwt. 89 lb. to pounds.
8. 26 pounds 3 ounces to ounces.
9. 2 short tons 6 cwt. to pounds.
10. 1 long ton to ounces.

Reduce:

11. 18 oz. to pounds and ounces.
12. 48 oz. to pounds and ounces.
13. 800 lb. to cwt.
14. 52 oz. to pounds and ounces.
15. 81 oz. to pounds and ounces.
16. 1200 lb. to cwt.
17. 118 oz. to pounds.
18. 2848 lb. to cwt. and pounds.
19. 8463 lb. to tons, cwt., and pounds.
20. 7861 oz. to higher denominations.

Liquid Measure.

1. How many times do you have to fill the pint measure to make one quart? To make one gallon?

2. How many quarts in 2 gallons? In 5 gallons? How many pints in 2 gallons?

3. What part of a quart is 1 pint? What part of a gallon is 1 pint?

4. Name liquids sold by the pint; by the quart; by the gallon.

Table of Liquid Measure.

2 pints (pt.)	=1 quart (qt.)
4 quarts	=1 gallon (gal.)

1. Mother canned 24 gal. of peaches in quart jars. How many jars did she fill?

2. A grocer bought 98 gal. of maple syrup and put it into quart bottles. How many bottles did he have?

3. He had 112 pint bottles of olive oil in stock. How many gal. orders could he fill with these?

WRITTEN EXERCISE

Reduce:

1. 8 qt. to pints.
2. 3 qt. 1 pt. to pints.
3. 4 gal. 3 qt. to quarts.
4. 7 gal. 2 qt. to pints.
5. 9 gal. 1 qt. 1 pt. to pints.
6. How many pints in 7 qt. 1 pt.?

7. How many pints in 34 gal.?
8. How many qt. in 59 gal.?
9. How many qt. in 143 gal.?
10. How many qt. in 72 gal. 3 qt.?

Reduce:

11. 32 pt. to quarts.
12. 48 pt. to quarts.
13. 11 pt. to quarts and pints.
14. 19 pt. to quarts and pints.
15. 80 pt. to gallons, quarts, and pints.
16. 126 qt. to gallons and quarts.
17. 157 qt. to gallons and quarts.
18. 700 pt. to higher denominations.

Dry Measure.

1. A pint is what part of a quart? How many pints in 4 qt.?

2. A quart is what part of a peck? How many quarts in 7 pk.?

3. A peck is what part of a bushel? How many pecks in 5 bu.?

4. A pint is what part of a peck? A quart is what part of a bushel?

Table of Dry Measure.

2 pints (pt.)	=1 quart (qt.)
8 quarts (qt.)	=1 peck (pk.)
4 pecks	=1 bushel (bu.)

1. Mother pays 18¢ a quart for cranberries. How much would a peck cost? How much would a bushel cost?

2. How many peck baskets of plums can a grocer fill from 25 bushels?

3. If he sells them in quart boxes, how many boxes will he need?

WRITTEN EXERCISE

Reduce:

1. 5 qt. to pints.

2. 10 qt. 1 pt. to pints.

3. 8 pk. to quarts.

4. 24 pk. 7 qt. to quarts.

5. 13 bu. to pecks.

6. 36 bu. 3 pk. to pecks.

7. 3 pk. 6 qt. to pints.

8. 8 bu. 2 pk. to quarts.

9. 10 bu. 1 qt. 1 pt. to pints.

10. 22 bu. 1 pk. to pints.

Reduce:

11. 10 pt. to quarts and pints.

12. 19 pt. to quarts and pints.

13. 48 qt. to pecks.

14. 41 qt. to pecks and quarts.

15. 24 pk. to bushels.

16. 71 pk. to bushels and pecks

Time Measure.

1. How many seconds in 3 minutes? How many minutes in 2 hours? How many hours in 2 days?

2. Name the months of the year.

3. What is a leap year?

Table of Time Measure.

60 seconds (sec.)	=1 minute (min.)
60 minutes	=1 hour (hr.)
24 hours	=1 day (da.)
7 days	=1 week (wk.)
4 weeks (about)	=1 month (mo.)
365 days	=1 year (yr.)
366 days	=1 leap year
12 months	=1 year
100 years	=1 century

How many:

1. Seconds in 34 min.?
2. Minutes in 7 hr.?
3. Hours in 3 da.?
4. Months in 6 yr.?
5. Hours in 640 min.?
6. Years in 82 mo.?
7. Hours in 1 wk. 3 da. 4 hr.?
8. Days in 312 hr.?

Measure of Counting.

12 things	=1 dozen (doz.)
12 dozen	=1 gross
20 things	=1 score

1. Mention five things sold by the dozen. How many oranges in 7 dozen?
2. How many pens in 3 gross? In 5 gross and 7?
3. How many years are 4 score and 10?
4. How many gross are 432 pencils?

The Thermometer.

The thermometer is marked off in **degrees** to tell the temperature of the air. As the air becomes warmer, the liquid in the tube rises; as the air grows cooler, the liquid falls.

1. Tell the temperature as shown by your thermometer.

2. Tell the temperature as shown in the picture.

3. Find the temperature at which water freezes. This is 32 degrees, or 32°.

4. Find the zero point on your thermometer.

5. How many degrees above zero is the temperature?

6. Water boils at 212°. Tell the number of degrees between the freezing point and the boiling point; between the present temperature and the boiling point.

7. At 6 A.M. the temperature was 61°; at 2 P.M. it was 76°. How many degrees did it rise?

8. At 11 P.M. it was 57°. How many degrees did it fall?

Paper Measure.

24 sheets	= 1 quire
20 quire	= 1 ream

1. How many sheets of note paper in 3 quires? In 10 quires?

2. How many quires in 3 reams? How many sheets in 3 reams?

3. How many sheets in 11 reams and 3 quires?

4. How many reams in 1440 sheets?

REVIEW PROBLEMS

1. How many one-ounce packages of pepper can the grocer make from $7\frac{1}{2}$ lb.?

2. Find the cost of $2\frac{1}{2}$ lb. of butter at 42¢ a lb.

3. How many ounces in $2\frac{1}{2}$ lb. of coffee?

4. The street is 72 ft. wide. How many yd. wide is it? How many inches in its width?

5. How many one-pound packages of coffee can a grocer put up from 228 oz.?

6. The fruit man sold 7 doz. peaches at 3¢ each. How much did he receive for them?

7. Mother used to pay 8¢ a qt. for potatoes. Now she buys them for 60¢ a pk. How much does she save on 7 pk.?

8. 5 tons of coal are sold for $33.75. How much would 60 tons cost at this rate?

9. A box of 12 doz. pens costs $.24. What would 180 doz. pens cost at this rate?

10. A farmer received $1551 for a shipment of hay. If the price was $22 a ton, how many tons did he ship?

11. At $6.85 a T., how much will 14 T. of coal cost?

12. A helper received 48¢ a bushel for picking blackberries. If he picked 37 bushels, how much did he earn?

13. How many glasses of milk can be sold from 6 gallons, if each glass holds one-half pint?

14. How many pieces of rope, each piece one yard long, can be cut from a coil containing 1137 feet?

15. How many bottles, each containing 1 pint of olive oil, can a dealer fill from a shipment of 128 gallons?

16. If a housekeeper pays $2.07 for 9 pounds of beef, how much will 15 pounds cost her?

17. A restaurant owner ordered 43 quart bottles of milk and 35 pint bottles of milk. Find the amount of his bill if milk is 10¢ a quart.

18. How many ounces in a package of food weighing $3\frac{1}{4}$ pounds?

19. A grocer paid $17.28 for 6 doz. bottles of strained honey. How much will he gain by selling it at $.35 a bottle?

20. A car conductor receives $2.35 a day. If a company employs 246 conductors, find the amount it pays in wages each day.

21. A girl bought a typewriter for $68.50, paying $13.50 in cash and the balance at the rate of $1.25 a week. How many weeks did it take her to pay for the machine?

22. How much greater is $\frac{6}{7}$ of 917 than $\frac{3}{5}$ of 990?

ORAL DRILL EXERCISE

	A	B	C	D
State results:				
1.	304×2	26+95	76×7	82−19
2.	506×8	44+76	56×8	54−28
3.	412×3	602×8	52×9	313×3
4.	67+33	81×9	67+19	216×2
5.	54+88	404×4	54+83	505×5
Divide:				
6.	2)1400	6)3000	2)2400	5)3500
7.	4)2400	5)2000	3)3600	6)1200
Find:				
8.	$\frac{1}{3}$ of 639	$\frac{1}{2}$ of 884	$\frac{2}{3}$ of 1200	$\frac{3}{5}$ of 300
9.	$\frac{1}{4}$ of 1212	$\frac{1}{5}$ of 500	$\frac{3}{4}$ of 800	$\frac{1}{2}$ of 900
10.	$\frac{7}{8}$ of 240	$\frac{2}{5}$ of 600	$\frac{1}{2}$ of 4800	$\frac{1}{8}$ of 3200
Tell the answers:				
11.	8 pk. = ? qt.	6×81	56 qt. = ? pk.	9×91
12.	$\frac{1}{2}$ gal. = ? qt.	77÷7	32 pk. = ? bu.	110÷10
13.	5×21	480÷8	20 qt. = ? gal.	340÷17
14.	758÷10	11 pk. = ? qt.	$\frac{3}{4}$ lb. = ? oz.	720÷12
15.	20 bu. = ? pk.	99÷9	$3\frac{1}{3}$ yd. = ? ft.	$280×10
16.	2×18	3×14	81×7	31×8
17.	72×4	12×7	35×200	310×30
18.	22×400	2×23	50×9	110×9
19.	410×5	62×4	3×13	84÷12
20.	96÷12	132÷12	5×14	132÷12

GENERAL ORAL PROBLEMS

1. A dealer paid $24 for 8 hats. He sold them for $4.75 each. How much did he make on each hat?

2. Our rent is $28 a month. Gertrude's father pays $39 a month for rent. How much more does he pay than we?

3. Harry paid 10¢ for 3 oranges. How much would he pay for 9 oranges? How much would 1 doz. cost?

4. Ethel bought 2 doz. eggs and used $\frac{1}{4}$ of them. How many did she use?

5. If the eggs sold for $.35 a doz., how much change did she receive from a two-dollar bill?

6. Cherries are $.12 a box. How many boxes can Ethel buy for $.75. How much change will she receive?

7. A dealer paid $31 each for 3 overcoats. How much change did he receive from $100?

8. He had 32 suits of clothes in his store. He sold $\frac{3}{4}$ of them. How many has he now?

9. He hopes to make $7 profit on each one of these remaining suits. How much will he make on all?

10. We had 24 drawing models in our closet. Miss Allen borrowed $\frac{5}{6}$ of them. How many did she borrow?

11. We never used more than $\frac{1}{3}$ of them at one time. How many was that?

12. We have 35 paint brushes. We are going to lend Miss Allen $\frac{3}{5}$ of them. How many will she have?

13. If 4 boxes of crackers cost a quarter, what will 12 boxes cost?

14. A lady bought 12 two-ounce boxes of cinnamon. How many pounds did she buy?

15. Take $\frac{1}{8}$ of 96 from $\frac{1}{5}$ of 200.

GENERAL WRITTEN PROBLEMS

1. Joe had a party. He ordered 2 gallons of ice cream at $1.35 a gal., a cake for 48¢, and a 2-pound box of candy at 40¢ a pound. How much did the party cost?

2. A class of 27 girls is going on a trip. How much will the tickets cost at $.95 apiece?

3. A dealer had 73 horses in one stable and 64 in another. He sold 34 from each stable. How many had he left?

4. Last week he bought 22 horses at $127 each. He sold them to the army at $141 each. How much did he gain?

5. A merchant bought 168 barrels of flour for $672. At this rate how many barrels could he buy for $984?

6. I paid $176 for a horse and when I sold it I gained $33.50. How much did I get for it?

7. Pads are sold for $.75 a dozen. How much will an order for 864 pads cost?

8. A man spent $\frac{1}{5}$ of his salary of $1500 for rent and $\frac{3}{5}$ of it for other expenses. How much did he have left?

9. His brother spent $\frac{3}{8}$ of $168 for pictures that cost $7 each. How many pictures did he buy?

10. A motorman received $3.45 a day. How much money will the company pay out to 12 motormen at the end of 24 days?

11. A street car makes its trip in 20 minutes. How many trips can it make in $4\frac{1}{3}$ hours?

12. A farmer gets $.18 a bunch for celery. He raised 516 stalks and tied them up, 4 stalks in a bunch. How much money did he receive?

13. If $\frac{3}{4}$ of this celery was sold for $.24 a bunch and the remainder at $.21 a bunch, how much was received for it?

14. The fruit man bought 176 boxes of cherries and sold $\frac{7}{8}$ of them. How many has he left?

15. He made 3¢ profit on each box he sold. Find his profit.

16. A man having $4674 in the bank drew out $\frac{5}{6}$ of it. How much was left in the bank?

17. A boy competed in a 660-yard dash and a 440-yard dash. How many feet did he run in both races?

18. Penknives cost 48¢ each. Find the cost of 8 dozen.

19. A dealer bought 18 desks at $16.45 each. He sold them and gained $8.25 on each desk. How much did he receive for them all?

20. If seven bicycles cost $147, find the cost of a dozen bicycles at the same price.

21. At 15¢ each, find the cost of 25 dozen napkins.

22. A dealer has 5472 inches of wire. If he cuts the wire into lengths, each one yard long, how many pieces does he cut?

23. A shopper bought a pair of rubbers at $1.15, a shawl at $2.50, 8 yards of cotton goods at $.21. How much did she have left of $10?

24. Find the cost of 4 pecks of apples at $.16 a quart, and $2\frac{1}{2}$ pounds of tea at 48¢ a pound.

25. A farmer bought an automobile for $540. When he sold it, he gained $\frac{2}{5}$ of what he paid for it. How much did he receive for it?

26. A school used 438 T. of coal in one year. Find the cost of this at $6.20 a T.

27. If the average expenses of running a large automobile truck are $18 a day, find the yearly expenses (counting 308 working days to the year).

28. How much greater is $\frac{4}{9}$ of 855 than $\frac{3}{8}$ of 944?

29. A grocer ordered Hawaiian pineapple at $2.10 a doz. cans, paying $56.70 for the shipment. How many cans did he order?

30. If silk costs $1.20 a yard, what will $\frac{7}{8}$ of a yard cost?

31. A suit manufacturer made 764 ladies' suits in six months. If he sold them at $8.35 each, how much did he receive?

32. If 245 five-pound bags of flour cost $75.95, how much will 300 bags cost?

ARITHMETIC BY GRADES

FOURTH YEAR BOOK

SECOND HALF: GRADE 4B

I. READING AND WRITING NUMBERS

Orders and Periods. Read the number:

548,926

This number shows six **orders**: *units,* shown by the first figure at the right, 6; *tens,* by the next figure, 2; then *hundreds,* shown by the figure 9; *thousands,* by the figure 8; *ten-thousands,* by the figure 4; and *hundred-thousands,* shown by the figure 5.

The seventh order (as in the number 1,000,000) is called *millions.*

The right-hand group of three figures, 926, is called **units' period.** The second group, 548, is called **thousands' period.** The periods are usually marked off by commas.

ORAL EXERCISE

1. Count by 100,000's from 100,000 to 900,000.
2. Read:

98,241	513,100	749,710	1,000,000
425,000	675,080	812,920	987,100
$4,812.50	$15,000	$22,463	$37,840.10

WRITTEN EXERCISE

Write in figures:

1. Eighty-four hundred nineteen.
2. One hundred sixty-thousand, seven hundred.
3. Five hundred thousand, one hundred.
4. Eight hundred fifty-four thousand, one hundred.
5. Seven hundred two thousand, nine hundred.

ORAL EXERCISE: ADVANCED WORK

1. Count by millions from 1,000,000 to 10,000,000.
2. Read:

1,200,000	7,710,000	9,463,025
9,621,800	4,001,000	12,637,914
3,650,000	8,000,000	10,000,000

WRITTEN EXERCISE: ADVANCED WORK

Write in figures:

1. One million, one hundred thousand.
2. Five million, nine hundred one thousand, six.
3. Six million, two hundred thousand, six hundred.
4. Seven million, fifteen.
5. Four million, one hundred ten thousand, forty.
6. Eight million, nine hundred thousand, twenty.

Roman Numerals.

1. Read: XXXVI LXXI LXXXVI XVI
XLVIII XCVIII XIV XL

2. Learn the following numbers:

C=100	CCC=300	D=500	DCC=700	CM= 900
CC=200	CD=400	DC=600	DCCC=800	M=1000

3. Tell three uses we make of Roman numbers.

Review of Roman Numbers. In the Roman notation, seven letters are used.

These letters are:	I	V	X	L	C	D	M
Their values are:	1	5	10	50	100	500	1000

The tens are written:

X XX XXX XL L LX LXX LXXX XC

The hundreds are written:

C CC CCC CD D DC DCC DCCC CM

ORAL EXERCISE

Read:

1. CCL	CDXX	DCXX	DCCCXLII
2. DCLXXX	CCCLX	DLXV	DCCXXXV
3. CMXI	CMLIX	CLXXXI	DLXIV

WRITTEN EXERCISE

Write in Roman numbers:

1. 120	500	880	990	250
2. 340	725	912	1000	480
3. 1917	1918	1919	1920	1921

II. COUNTING

I. Count:

1. By 2's from 7 to 37.
2. By 3's from 5 to 41.
3. By 4's from 7 to 55.
4. By 5's from 6 to 56.
5. By 6's from 5 to 77.
6. By 7's from 4 to 88.
7. By 7's from 3 to 80.
8. By 8's from 3 to 83.
9. By 9's from 2 to 92.
10. By 9's from 7 to 79.

II. Count backward:

1. By 3's from 35 to 2.
2. By 4's from 51 to 3.
3. By 5's from 66 to 6.
4. By 6's from 79 to 7.
5. By 6's from 74 to 8.
6. By 7's from 87 to 10.
7. By 5's from 71 to 21.
8. By 10's from 92 to 12.
9. By 8's from 95 to 15.
10. By 7's from 78 to 1.

III.

1. Beginning with 8, count by 3's to 68.
2. Beginning with 4, count by 7's to 53.
3. Beginning with 3, count by 2's to 41.
4. Beginning with 9, count by 8's to 97.
5. Beginning with 11, count by 6's to 83.
6. Beginning with 1, count by 7's to 78.

IV.

1. From 87, count by 5's to 7.
2. From 91, count by 4's to 3.
3. From 112, count by 10's to 2.
4. From 136, count by 9's to 10.
5. From 103, count by 7's to 12.
6. From 95, count by 9's to 5.

III. ADDITION

ORAL EXERCISE

Give sums rapidly:

1.

$1.80	$.30	$.10	$.50	$.20	$.70	$.40	$.90
.20	.70	.50	.40	.20	.80	.30	.60
.60	.50	.60	.70	.50	.40	.70	.80

2.

56	17	35	72	69	44	27	81
27	48	34	18	23	37	53	19

3.

11	83	29	61	57	34	76	45
49	13	29	37	36	49	18	39

Add:

4.

50	70	20	10	40	30	90	60
30	10	80	20	50	60	30	40
25	39	69	75	35	49	29	18

5. In addition what are the numbers to be added called?

6. What is the result called?

Add:

7.

88¢	39¢	69¢	25¢	75¢	50¢	49¢	19¢
42¢	39¢	75¢	82¢	36¢	45¢	49¢	36¢

8.

$ .11	$ .17	$ .14	$ .16	$ .30	$ .15
.90	.50	.60	.40	.80	.30
2.00	6.00	7.00	3.00	5.00	8.00

WRITTEN EXERCISE

Add and check:

1.	2.	3.	4.
29,364	$632.45	30,321	$960.68
96,068	965.80	22,121	400.22
98,420	242.63	42,807	362.45
57,943	866.44	9,607	817.61
3,014	967.87	8,976	608.03
65,400	50.00	14,803	273.21
31,533	14.72	54,394	186.39
60,803	654.00	5,433	242.65

5.	6.	7.	8.
93,150	$630.20	111,805	$763.75
33,085	776.30	112,514	971.28
86,656	992.42	186,008	994.24
96,025	134.43	54,191	873.60
8,105	549.20	6,034	717.50
78,066	53.69	54,672	890.44
65,119	50.56	62,150	99.68
14,360	394.05	12,235	858.73

9.	10.	11.	12.
39,416	$665.20	28,901	$668.89
57,632	205.00	92,837	526.08
5,643	345.52	93,190	161.03
69,837	883.63	2,913	271.13
38,641	43.71	57,926	428.75
32,928	843.81	60,256	324.43
3,209	430.44	28,706	761.25
45,411	300.52	8,091	943.28

ORAL DRILLS

	1	2	3	4	5	6	7
a	4	13	24	31	45	50	60
b	8	17	26	33	46	56	62
c	9	11	22	35	41	57	67
d	3	14	25	36	47	53	65
e	7	12	23	34	44	55	63
f	5	18	27	32	43	54	61

1. Adding down by columns, add one of the following numbers to each column of numbers in the preceding drill chart: 62; 41; 26; 35; 70; 56.

2. Adding across by rows, add one of the following numbers to each row of numbers in the above drill chart: 53; 29; 81; 72; 36; 45.

3. Add the following columns as rapidly as possible. As you add look for groups of two or more numbers that make 10.

	1	2	3	4	5	6	7	8	9	10	11	12
a	2	8	7	2	5	7	2	5	5	2	7	9
b	8	7	9	8	7	5	8	9	5	9	3	3
c	4	3	6	7	9	6	7	7	9	4	6	4
d	9	4	5	9	2	4	6	3	8	6	4	7
e	7	7	3	8	8	9	8	8	4	7	9	8
f	3	7	2	5	7	8	5	2	5	5	8	7
g	5	9	8	6	8	9	9	7	6	3	9	3
h	4	2	9	3	1	6	7	8	8	2	8	6
i	1	8	7	7	9	4	5	7	8	2	9	2

4. In the preceding chart, add the numbers across by rows. Check by adding in the opposite direction.

WRITTEN EXERCISE

Add and test:

1.	2.	3.	4.
21,204	$397.82	48,976	$438.76
78,816	365.48	46,543	246.66
10,576	432.10	8,610	420.60
31,681	52.87	76,498	90.90
26,081	784.90	99,327	51.63
8,536	896.38	78,000	92.14
56,790	976.62	69,053	27.65
84,605	783.81	3,499	853.48

5.	6.	7.	8.
45,088	21,342	35,353	53,683
57,028	18,173	1,807	7,845
9,646	78,470	5,480	54,790
35,467	3,638	20,006	78,966
57,000	25,942	34,678	69,000
16,170	6,746	19,843	87,211
8,150	12,345	8,467	72,720
19,512	9,031	25,244	8,654

9.	10.	11.	12.
$476.98	75,683	$460.16	75,268
495.98	49,678	917.28	93,876
356.40	3,876	649.26	68,097
676.79	83,597	265.34	41,863
467.65	20,676	834.77	46,875
937.65	54,978	720.00	89,790
220.00	18,392	834.77	7,386
186.93	93,780	546.70	54,398

WRITTEN REVIEW PROBLEMS

1. On Monday 894 people attended the ball game, on Tuesday 2,307, Wednesday 3,291, Thursday 2,565, Friday 1,875, Saturday 5,068. How many people attended during the week?

2. A steamer on a 6-day trip made the following distances daily: first day 439 miles, second 516 miles, third 528, fourth 521, fifth 534, sixth 443. How far did she travel?

3. A man bought a farm for $8,760. He built a house costing $4,750 and a barn worth $1,955. He sold the whole farm for $21,250. How much did he gain?

4. A merchant put the following amounts in a bank: $5,473.25, $8,917.68, $3,004.82, $7,675.83, $4,564.16 and $6,017.23. How much did he have left in the bank after he drew out $11,267.23?

5. Eight coal barges were tied up at a dock. The first contained 3,540 tons of coal, the second 4,115 tons, the third 3,337 tons, the fourth 3,516 tons, the fifth 1,813 tons, the sixth 2,646 tons, the seventh 1,997 tons, the eighth 3,776 tons. Find the total amount they contained.

6. A large store took in $11,000 in a week. On Monday it received $1,875.20, on Tuesday $963.40, Wednesday $1,246.32, Thursday $1,163.38, Friday $2,135.46. Find how much was taken in on Saturday.

IV. SUBTRACTION

ORAL EXERCISE

Tell the difference:

1.	73¢	97¢	46¢	52¢	21¢	34	63	81
	48¢	85¢	29¢	37¢	11¢	19	57	62
2.	42¢	97¢	38¢	54¢	80¢	23¢	71¢	73¢
	27¢	89¢	26¢	47¢	55¢	11¢	58¢	45¢

3. Tell which numbers are the minuends in the above examples. Which are the subtrahends?

4. What is the result called in subtraction?

5. Tell how we check work in subtraction.

Give answers:

6.	50+20−40	80+20−30	10+70−40	60+40−30
7.	60+10−40	90+10−50	30+30−20	70+20−60
8.	42+20−8	25+30−9	47+20−5	63+20−8
9.	31+40−8	16+40−6	54+30−7	29+50−9

10. Tell how much change a clerk should give out if he is handed a dollar bill and the purchases are the following:

	1	2	3	4	5	6	7	8
a	72¢	65¢	77¢	97¢	84¢	54¢	91¢	88¢
b	59¢	85¢	51¢	81¢	68¢	87¢	66¢	93¢
c	35¢	15¢	65¢	25¢	45¢	85¢	55¢	75¢
d	47¢	24¢	76¢	36¢	67¢	11¢	27¢	18¢

WRITTEN EXERCISE

Subtract and check:

1. 9751 8875	**2.** $347.02 54.66	**3.** 96,541 78,782	**4.** $899.00 569.83
5. 84,633 69,887	**6.** 500,641 495,586	**7.** 766,103 59,749	**8.** 484,311 58,499
9. $9382.01 8987.63	**10.** 721,470 78,685	**11.** 267,543 198,785	**12.** $6449.00 586.97
13. 436,900 434,897	**14.** $8400.00 768.26	**15.** $5367.11 943.29	**16.** 743,611 589,743

Subtract:

17. 700,000 − 246,843

18. $9,243.18 − $439.79

19. $74,281 − $35,986

20. 835,040 − 747,395

21. 335,000 − 296,683

22. 167,201 − 98,563

23. 10,008 − 4,598

24. $2237.11 − $1948.65

ORAL EXERCISE

Give the answers in order:

1. From 76 take:	26	37	48	53	58	44	69
2. From 92 take:	18	27	34	46	57	66	83
3. From 84 take:	73	65	47	36	28	19	55

4.	**5.**	**6.**	**7.**
30+10−20	70+20−30	20+50−10	60+30−40
60+10−30	80+10−50	30+40−20	70+20−50
48+10−30	32+20−50	27+30−20	50+20−40

8. State the amount of change received in each of these purchases:

	a	*b*	*c*	*d*	*e*	*f*	*g*	*h*
$1.00	$ .45	$ .55	$ .62	$ .38	$ .19	$ .75	$ .82	$ .33
	.65	.36	.11	.41	.87	.21	.49	.54
$2.00	1.70	1.35	1.78	1.22	1.76	1.54	1.65	1.27
	1.18	1.69	1.45	1.61	1.48	1.89	1.91	1.38
$5.00	2.20	4.68	2.38	3.14	3.50	2.22	3.45	4.18
	2.40	4.91	3.10	2.62	4.18	2.87	4.75	3.15
$10.00	7.15	5.50	6.40	5.20	8.30	5.45	7.50	5.75
	8.40	5.10	7.30	6.80	7.65	8.95	9.10	6.20

Give answers:

9.	**10.**	**11.**	**12.**
35+40−14	63+30−13	72+20−12	34+30−14
78+20−16	69+20−18	47+30−17	28+50−18
86+10−14	46+40−15	39+50−19	77+20−16
48+20−17	35+40−15	42+40−15	48+10−17

WRITTEN EXERCISE

Subtract and prove:

1. 304,700 8,946	**2.** $1731. 1493.65	**3.** 232,512 48,987	**4.** $5647. 4987.18
5. 121,830 119,946	**6.** 637,500 1,496	**7.** $1733. 45.86	**8.** 594,406 94,708
9. 609,543 86,777	**10.** 273,501 69,612	**11.** 348,762 98,884	**12.** $2847.12 689.55

WRITTEN REVIEW PROBLEMS

1. The subtrahend is 164,387; the minuend is 204,133. Find the difference.

2. In an army of 307,100 men, 28,486 were wounded in battle. How many were unwounded?

3. An agent bought a farm for $8,462.50. He sold it for $10,150. How much was his gain?

4. From twenty-one thousand, seven hundred sixty-three, subtract eleven thousand, nine hundred eighty.

5. The population of a city was 797,883. In ten years it increased to 1,116,715. What was the increase in ten years?

6. A store sold $21,817.61 worth of goods in six months. The expenses were $16,998.75. How much were the profits?

7. A company placed in a bank the following sums: $3,487.62, $18,267.17, $9,008.75, $13,176.83, and $22,174.78. If the company drew out $25,864.16, how much still remains in the bank?

8. The area of one state is 18,618 sq. mi. The area of another is 31,105 sq. mi. Find the difference in area.

9. The expenses of running a coal mine were $12,460.50 in March, $10,288.84 in April, and $11,057.70 in May. Find the total expense for the three months.

10. A company bought a plot of land for $18,273.40, erected a hotel at a cost of $22,100.20, and purchased furniture worth $6,588.05. How much money did the company spend?

ORAL DRILL EXERCISE

	A	B	C	D
	State the products:			
1.	62×7	95×8	63×9	71×8
2.	86×9	75×8	93×7	48×5
3.	23×8	34×7	45×5	56×6
	Tell the answers:			
4.	$56+30-17$	$34+40-11$	$76+20-15$	$32+50-17$
5.	$32+14-7$	$41+22-2$	$54+25-18$	$63+19-16$
6.	$84+26-14$	$97+17-8$	$18+28-9$	$26+29-10$
7.	$14+21-20$	$45+52-19$	$32+18-21$	$56+31-17$
	Find quotients:			
8.	80)640	30)4500	60)720	70)840
9.	$3600 \div 40$	$450 \div 50$	$1400 \div 20$	$1800 \div 90$
10.	$1800 \div 20$	$7000 \div 10$	$810 \div 90$	$1590 \div 30$
11.	50)3500	60)4200	80)720	50)3500
	Give results:			
12.	9×12	10×12	12×40	12×60
13.	7×12	12×80	12×20	6×12
14.	$120 \div 12$	$720 \div 12$	$480 \div 12$	$960 \div 12$
15.	$240 \div 12$	$600 \div 12$	$360 \div 12$	$840 \div 12$

GENERAL ORAL PROBLEMS

1. Helen bought $1\frac{1}{2}$ lb. fish at 12¢ a lb. How much did she pay for it?

2. If 8 peaches cost 24¢, what will a dozen peaches cost?

3. We have 45 pupils in our class; $\frac{4}{5}$ are going to our Field Day. How many will not attend?

4. Rice is 9¢ a pound. What is the cost of 3 lb.? 8 lb.? 11 lb.? $\frac{1}{2}$ lb.? $\frac{1}{3}$ lb.? $\frac{2}{3}$ lb.? $\frac{1}{4}$ lb.? $\frac{3}{4}$ lb.?

5. Mary had 8 yd. of ribbon. If she used $1\frac{3}{4}$ yd., how many yd. were left?

6. Last week she had 59 yd. and her mother bought 63 yd. more. How many had they both?

7. If 3 lb. of baking powder cost one dollar, find the cost of 15 lb.

8. Mother spent $\frac{3}{4}$ of an hour in the market and $1\frac{2}{3}$ hours at a friend's house. How long was she away?

9. If 3 eggs cost 12¢, what will $\frac{1}{2}$ a dozen cost?

10. Ethel bought 4 yd. of ribbon at 72¢ a yard. Find the cost.

11. How much change should she get from a two-dollar bill?

12. Butter is 32¢ a lb. How much must I pay for $1\frac{1}{2}$ lb.?

13. If you spend $\frac{3}{8}$ of 48¢, how much money have you left?

14. If a bushel of oats weighs 32 pounds, what will a peck of oats weigh?

15. Find the weight of 50 bushels of oats at this rate.

16. A lady bought a pair of shoes at $9.40 and an umbrella at $2.50. How much change will she receive from $20?

17. If our newspapers cost 48¢ a month, how much will they cost for a year?

18. Find the cost of 120 shirts at $1.50 each.

GENERAL WRITTEN PROBLEMS

1. How many one-ounce cans can the grocer fill from 50 pounds of pepper?

2. From $5, how many cents will Helen have left if she buys 11 yd. of silk at 45¢ a yd.?

3. An agent sold 11 sewing machines for $693. What would he receive for 2 doz. machines at this rate?

4. By running errands, John earns a quarter every school day and $.75 on Saturday. How much does he earn in a week?

5. A lamp costs $.85. It is sold at $1.25. Find the profit on the sale of 4 doz. lamps.

6. 13 doz. pairs of shoes were bought for $663. Find the cost of one pair.

7. A grocer bought 36 gallons of oil for $59.40. He sold it at 24¢ a pint. Find his gain.

8. $1.75 each was paid for 26 copies of a book. They were all sold for $65. How much was gained on each copy?

9. A woman bought a sewing machine for $65, paying $30 in cash and the balance at the rate of $.50 a week. How many weeks did it take her to pay for the machine?

10. A delivery truck travels 94 miles the first day, 135 miles the second day, 84 miles the third day, and 72 miles the fourth day. Find the average distance traveled each day.

11. A dealer bought 85 pictures for $272 and sold them at $3.25 each. How much did he gain?

12. Find the cost of 36 parlor lamps at $15.25 each.

13. The total number of passengers and crew on a steamer was 2,256 people. There were 211 first-class passengers, 463 second class, and 839 third-class passengers. Find how many people in the crew of the ship.

14. We have 2,347 pupils in our building. 721 are on the first and second floors and 883 are on the third floor. How many are on the top floor?

15. A news dealer bought 7 doz. magazines at 11¢ each and sold them at 15¢ each. How much did he gain?

16. Find the cost of 25 gallons of olive oil at 45¢ a pint.

17. If a workman is paid 45¢ an hour for overtime, how much will he earn in 18 evenings, working from 7 o'clock to 9.30 each evening?

V. MULTIPLICATION

ORAL EXERCISE

Give answers rapidly:

1.	2.	3.	4.	5.	6.
6×7	7×9	5×6	9×5	4×13	4×16
8×5	9×9	6×9	7×8	5×14	5×15
6×8	9×6	9×4	7×12	7×13	6×18
9×7	6×20	7×6	8×9	4×12	5×9
7×7	9×11	6×12	5×12	5×16	6×16
8×12	3×14	9×12	4×11	6×15	4×25
6×11	8×8	7×11	5×20	5×11	9×8

7. Recite the table of 6's; of 8's; of 9's.

8. State the products in the table of 7's; of 11's; of 12's.

9. Multiply each by 2: 44, 37, 36, 19, 50, 48, 34.

10. Multiply each by 3: 33, 25, 31, 23, 17, 18, 15.

11. Multiply each by 5: 14, 13, 12, 18, 15, 20, 16.

12. Multiply each by 4: 22, 16, 13, 15, 24, 14, 20.

13. How many days in 7 wk.? 11 wk.? 15 wk.? 12 wk.? 20 wk.? 30 wk.? 13 wk.? 14 wk.?

14. How many quarts in 8 pk.? 12 pk.? 20 pk.? 25 pk.? 30 pk.? 15 pk.? 14 pk.? 18 pk.?

15. How many months in 8 yr.? 10 yr.? 20 yr.? 9 yr.? 11 yr.? 12 yr.? 13 yr.? 15 yr.?

WRITTEN EXERCISE

1. A farmer sold 84 chickens at $.80 each. How much did he receive?

2. A class uses 32 pen points every week. How many will they use in 25 weeks?

3. A grocer sold 28 bottles of syrup at $.75 each. How much did he receive for them?

4. I bought 18 packages of chocolate at $.22 each. How much did I pay?

5. An automobile uses a gallon of gasoline for every 17 miles. How far can it travel on 87 gal.?

6. Cocoa is 27¢ a can. Find the cost of 94 cans.

7. If an automobile goes 34 miles in an hour, how far can it travel in 17 hours?

8. A merchant sold 85 caps at $49¢ each. How much did he get for them?

9. These caps cost 31¢ each to make. How much did they cost?

ORAL EXERCISE

1. Flour is 4¢ a pound. Find the cost of 40 lb.; 25 lb.; 12 lb.; 31 lb.; 37 lb.; 35 lb.; 50 lb.; 61 lb.

2. Sugar is 6¢ a pound. Find the cost of 30 lb.; 100 lb.; 12 lb.; 15 lb.; 20 lb.; 40 lb.; 80 lb.; 200 lb.

3. Coal is $8 a ton. Find cost of 8 T.; 12 T.; 100 T.; 20 T.; 25 T.; 60 T.; 50 T.; 200 T.; 22 T.

4. Multiply each by 2: 14, 17, 23, 35, 44, 81, 90.

5. Multiply each by 3: 16, 24, 46, 51, 65, 80, 91.

WRITTEN EXERCISE

Multiply these examples. Remember that in multiplying dollars and cents, we point off two places in the answer.

1. $4.20 4	2. $8.35 5	3. $9.82 7	4. $16.20 7
5. $56.37 7	6. $39.32 8	7. $84.16 9	8. $90.52 6
9. $63.17 2	10. $54.32 7	11. $24.19 5	12. $36.12 4
13. $34.35 6	14. $46.70 8	15. $87.31 9	16. $90.16 4

17. $16.54×9	20. $73.18×25	23. $22.62×43
18. $56.71×13	21. $60.50×41	24. $18.94×37
19. $50.82×22	22. $936.14×4	25. $50.75×20

26. What do we call the number to be multiplied?

27. What name is given to the result in multiplication?

28. How can we check this result?

Multiply and check:

29. 133×86	33. 783×12	37. $1642×19
30. 876×67	34. 4246×9	38. 224×36
31. $4.89×28	35. 426×84	39. $8.12×36
32. $22.25×92	36. $16.22×93	40. 7563×44

ORAL EXERCISE

Give products, multiplying both ways:

1. 9×	11	9	12	22	51	60	×7	5.
2. 6×	12	20	14	15	17	30	×5	6.
3. 10×	100	56	300	95	89	48	×3	7.
4. 11×	40	7	20	11	70	30	×8	8.

9. How many feet in 20 yd.? 14 yd.? 31 yd.? 34 yd.? 13 yd.? 27 yd.? 33 yd.?

10. How many pecks in 21 bu.? 17 bu.? 11 bu.? 24 bu.? 16 bu.? 14 bu.? 19 bu.? 12 bu.?

11. Tell the cost of 5-cent fares for 13 people; for 12; for 15; for 18; for 16; for 14; for 9; for 17.

Multiplying by Multiples of 10.

860	$860	720	720	$5.25	$14.72
10	40	100	300	10	200
8600	$34400	72000	216000	$52.50	$2944.00

To multiply by 10, *add a zero or move the decimal point one place to the right.*

To multiply by 100, *add two zeros or move any decimal point two places to the right.*

$52.50×100=$5250.

To multiply by 20, 30, 40, *etc., add the zero and multiply by* 2, 3, 4, *etc.*

To multiply by 200, 300, 400, *etc., add two zeros and multiply by* 2, 3, 4, *etc.*

ORAL EXERCISE

Multiply:

1.	2.	3.	4.	5.
88×10	92×20	35×20	116×10	61×100
63×10	48×20	35×100	125×10	81×40
310×10	25×30	35×200	69×100	33×200
56×100	60×100	335×10	812×10	47×100

WRITTEN EXERCISE

Multiply:

1. 463×10
2. 724×100
3. $52.20×10
4. $1.48×10
5. 526×10
6. $18.40×100
7. $75×500
8. $1.48×100
9. 452×60
10. 631×50
11. 78×200
12. $1.48×50
13. 591×40
14. 89×300
15. $65.45×100

Find the products and check results:

16. $1.48×50
17. $65.14×83
18. $5.22×28
19. 593×10
20. 5934×100
21. $47.92×98
22. 583×700
23. $858×630
24. 1876×40
25. $73.21×100
26. 8439×20
27. 9647×170
28. 563×240
29. 269×800
30. 654×480
31. 7314×30
32. $73.75×69
33. 347×400
34. $8.55×75
35. 913×90
36. 565×360
37. $57.16×84
38. 5642×30
39. 347×900
40. 793×690
41. 807×79
42. $2.83×60
43. $17.28×40
44. $9.11×326
45. $91.11×30
46. $63.17×87
47. $54.25×220
48. 1847×480

WRITTEN EXERCISE

Find the cost of:

1. 20 chairs at $7.74 each.
2. 90 lb. butter at 39¢ a lb.
3. 6 doz. coats at $12.60 each.
4. 180 cases of eggs at $10.40 a case.
5. 240 sacks of flour at $.95 a sack.
6. 180 suits of clothes at $15.45 each.
7. 75 desks at $8.60 each.
8. 164 yd. carpet at $.97 and 73 yd. silk at $.83 a yd.
9. 480 lb. of coffee @ 38¢ a lb.
10. 8 doz. rosebushes @ 35¢ each.
11. 75 bottles mucilage @ 8¢ a bottle.
12. A storekeeper bought 35 tables for $400.75. He sold them for $16 each. Find his profit.
13. A man bought a house for $8460. He spent $375.48 on painting and $426.45 on plumbing repairs. He sold the house for $9250. Did he gain or lose, and how much?
14. If one parlor table costs $45.20, how much will 90 tables cost?
15. Mr. Snow's monthly expenses are $160.75. If he earns $225.50 a month, how much can he save in a year?
16. A steamship can travel 483 miles a day. At this rate, how many miles can it travel in 8 days?

Multiplying by Three Figures.

1. Multiply 284 by 420:

		CHECK
284 420 5680 1136 119280	Notice that we bring down the 0, and multiply by 42. We are careful to place the right-hand figure 6 of the second partial product under the multiplying figure 4.	420 284 1680 3360 840 119280

2. Multiply 347 by 186:

		CHECK
347 186 2082 2776 347 64542	Notice that we write the first partial product with the right-hand figure under the multiplier 6. We write the second partial product with the right-hand figure under the multiplier 8. And we write the third with the right-hand figure under 1.	186 347 1302 744 558 64542

3. Multiply 204 by 402:

		CHECK
204 402 408 8160 82008	Here we multiply first by 2. Then we bring down the 0 since the product by 0 tens is 0. Multiplying by 4, we place the 6 of the partial product under the 4.	402 204 1608 8040 82008

WRITTEN EXERCISE

Find the products:

1. 198×300	13. 340×118	25. 290×500
2. 325×170	14. \$3.26×400	26. \$1.28×407
3. \$6.67×307	15. 567×149	27. 156×328
4. 415×239	16. 687×102	28. 119×425
5. 235×128	17. 786×134	29. 483×608
6. 324×223	18. 345×120	30. \$6.10×86
7. 137×308	19. \$7.84×65	31. 176×205
8. 242×190	20. \$120×340	32. 235×320
9. 565×781	21. 745×336	33. \$634.12×23
10. \$82.56×43	22. 654×300	34. 534×507
11. 493×607	23. \$818×207	35. 386×500
12. \$564.13×18	24. 785×400	36. 686×708

WRITTEN EXERCISE

Multiply:

1. 246×123	13. 139×903	25. 196×642
2. 847×246	14. 643×406	26. 187×600
3. 363×374	15. 547×708	27. \$2.09×487
4. \$508×226	16. 991×609	28. 614×275
5. 712×765	17. 3468×96	29. 272×150
6. \$609×348	18. 8123×97	30. 532×319
7. 894×129	19. 4006×615	31. 272×150
8. \$343×931	20. 556×3004	32. \$5.16×72
9. 345×205	21. \$4398×165	33. 243×272
10. 623×304	22. \$10.12×808	34. \$354.63×17
11. 742×702	23. \$3,246×209	35. 643×300
12. 593×809	24. 1835×806	36. 4983×76

WRITTEN REVIEW PROBLEMS

1. A farmer sold 18 cows at $52 each and in payment took a 720-dollar automobile and cash. How much cash did he get?

2. Oranges sell at 3 for a quarter. How much will 13 doz. cost?

3. How much change will Helen receive from a five-dollar bill if she buys $14\frac{1}{3}$ lb. of beef at $24 a lb.?

4. I can hire a flat at $39.75 a month or I can take it for a year at $465. Which way is the cheaper and how much shall I save each month?

5. A dealer bought 2 dozen typewriters for $1158. He sold them for $55 each. How much did he gain?

6. A clerk received $279 for 9 weeks' work. How much would he get for 4 months' work at this rate?

7. 14 automobile tires were sold for $252. What would 30 tires cost at this price?

8. A stationer bought notebooks at 48¢ a doz. and sold them at 5¢ each. How much did he make on 14 dozen?

9. A dealer bought 58 pounds of coffee at $.24 a pound and sold it at $.32 a pound. How much did he gain?

10. Find the cost of 19 cases of jelly each containing $2\frac{1}{2}$ doz. jars at 4¢ each.

11. The amount taken in by a department store at the end of a week was $18,476.50. If the expenses were $\frac{5}{7}$ of this, find the amount of the expenses.

ORAL DRILL EXERCISE

A	B	C	D
Subtract:			
1. \$1−.35	\$2−1.08	\$3−1.60	\$4−2.20
2. \$1−.66	\$2−1.39	\$3−2.04	\$4−3.10
3. \$1−.81	\$2−1.45	\$3−2.15	\$4−3.30
4. \$1−.49	\$2−1.55	\$3−2.35	\$4−3.85
Give results:			
5. 39¢×10	23×8	35×20	66÷5
6. 45÷4	41÷3	49×4	42×40
7. \$5.20×10	71×30	45÷7	63÷5
8. 54×7	27÷8	\$2.57×100	37×6
Add:			
9. 58 42 12 74	67 13 49 63	75 27 29 85	89 39 91 19
0. 86 27 15 95	69 11 55 85	42 94 75 35	18 84 15 25
1. 92 42 33 96	22 52 28 79	82 12 46 37	32 72 81 57
Find:			
2. $\frac{3}{4}$ of 52	$\frac{1}{2}$ of 280	$\frac{2}{3}$ of 75	$\frac{3}{4}$ of 88
3. $\frac{1}{2}$ of 144	$\frac{1}{4}$ of 800	$\frac{3}{4}$ of 60	$\frac{2}{3}$ of 45
4. $\frac{2}{3}$ of 900	$\frac{1}{10}$ of 110	$\frac{1}{8}$ of 640	$\frac{1}{5}$ of 105

GENERAL ORAL PROBLEMS

1. A dealer sold 20 typewriters at \$62 each. How much did he receive for them?

2. If he pays $\$6\frac{1}{3}$ for 2 doz. typewriter ribbons, how much must he pay for 6 dozen?

3. He received $\$4\frac{3}{4}$ for repairing one machine and \$.50 for repairing another. How much did he receive for both?

4. A customer paid him \$8.35 for repairs and \$.45 for a ribbon. From a 10-dollar bill, how much change should the dealer give the customer?

5. New York City is 150 miles from Albany. After traveling $\frac{4}{5}$ of the way, how many more miles has a train to travel?

6. What will 2 bu. of potatoes cost at \$.05 a qt.?

7. If a doz. booklets cost \$3, what will 144 cost?

8. A tub of butter weighs $36\frac{3}{4}$ lb. The tub weighs $10\frac{1}{4}$ lb. How much does the butter weigh?

9. The difference between two prices is \$.38. The larger one is \$.75. What is the smaller?

10. Find the cost of 10 collars at two for 25¢.

11. How many feet in $4\frac{1}{2}$ yards of garden hose?

12. $\frac{1}{8}$ of the eggs in an order of 6 dozen were broken How many were broken?

13. Find the cost of 8 doz. erasers at 5¢ apiece.

14. If a boy earns \$28 a month and spends \$25.50 a month, how much can he save in a year?

GENERAL WRITTEN PROBLEMS

1. An agent bought 326 sheep at $4.25 each. On selling them, he gained $147.50. How much did he receive for them?

2. A dealer sold a carriage for $391, gaining $74.25. How much did he pay for it?

3. A man had $14,987 in a bank. He drew out $5275 and later put in $2730. How much had he in the bank then?

4. If 10 pairs of shoes cost $55, what will 2 doz. pairs cost?

5. Father bought shoes for $\$6\frac{1}{2}$, a coat for $\$17\frac{3}{4}$, and a vest for $\$5\frac{1}{4}$. How much did he pay in all?

6. 6 doz. schoolbooks cost $30.24. How much did one book cost?

7. If $9.60 is spent for 16 lb. of candy, find what 64 lb. will cost.

8. The rent of a store is $480 a year. Find the rent for $3\frac{1}{4}$ yr.

9. A dealer exchanged 70 bbl. of sugar @ $22.50 a barrel for flour @ $5 a barrel. How many barrels of flour did he receive?

10. A contractor hired 34 men at $2.15 a day for 21 days. How much did he have to pay them?

11. The population of a state is 856,353. $\frac{2}{3}$ of the people live in towns and cities. How many live on the farms?

12. A man having $10,420.08 in the bank paid bills amounting to $868.05, $1473.10, $950, $1728.86, and $989.95. How much has he in the bank now?

13. A train runs at an average speed of 34 miles an hour. How many hours will it take to make a trip of 2346 miles?

14. A storekeeper paid $240.80 for 14 ladies' suits. How many suits at the same price could he buy for $172?

15. Find the cost of 9 dozen tennis rackets at $2.40 each.

16. A farmer exchanges 72 dozen eggs at 25¢ a dozen, for shoes at $6 a pair. How many pairs can he get?

17. The storekeeper paid $5.15 a pair for these shoes. Find his profit on a case containing 3 dozen pairs.

18. A dealer paid a factory $12,028.16 for two shipments of furniture, paying $5897.58 for the first. How much did he pay for the second?

19. A merchant bought wheat at 88¢ a bushel and sold it at $1.07 a bushel. What was his profit on 1845 bushels?

20. A family uses 5 pt. of milk each day. If the milk costs 10¢ a qt., what will be the amount of their milk bill for the month of March?

21. A grocer, having purchased a barrel containing 196 pounds of flour, put up the flour in 7-pound bags. If he sold it at 36¢ a bag, how much did he receive for the contents of the barrel?

VI. DIVISION

Factors and Multiples.

$2 \times 3 = 6$	2 and 3 are the *factors* of 6.
$3 \times 2 \times 5 = 30$	3, 2, and 5 are the *factors* of 30.
$5 \times 3 \times 2 \times 7 = 210$	5, 3, 2, and 7 are the *factors* of 210.

The **factors** of a number *when multiplied together* will make up the number.

The multiplicand and the multiplier are the factors of the product.

$3 \times 5 = 15 \qquad 15 \div 3 = 5 \qquad 15 \div 5 = 3.$

An exact divisor of a number is a factor of. that number. If we have the product and one factor, we find the other factor by dividing.

$3 \times 2 = 6$	6 is called a *multiple* of 2 and cf 3.
$3 \times 4 = 12$	12 is called a *multiple* of 2, of 3, of 4, and of 6.
$2 \times 6 = 12$	

These numbers are called **multiples** because they are *exactly divisible by the factors.*

ORAL EXERCISE

1. Name two factors of each of these numbers:

64	21	39	46	50	88	121	100
96	78	70	60	84	54	138	144

2. Tell the first twelve multiples of 3; of 4; of 5; of 7; of 9; of 11; of 8; of 6.

The left-hand number in each of the following examples is a factor of every number in its row. Find the other factor of each number.

3.	2	32	50	36	42	88	100	18
4.	3	45	33	27	48	9	60	30
5.	4	28	48	88	36	44	100	52
6.	5	40	85	95	55	100	200	125
7.	7	49	70	98	84	140	91	42
8.	8	48	96	88	400	104	120	800
9.	9	81	108	540	360	900	450	810

Supply the missing multiples and factors:

10.	11.	12.	13.
$? \div 3 = 21$	$70 \div ? = 14$	$? \div 6 = 30$	$? \div 2 = 36$
$108 \div ? = 54$	$? \div 16 = 6$	$90 \div ? = 45$	$? \div 4 = 16$
$? \div 5 = 18$	$? \div 3 = 19$	$? \div 7 = 14$	$96 \div ? = 8$
$84 \div ? = 3$	$? \div 6 = 14$	$39 \div ? = 3$	$360 \div ? = 6$

Supply the missing factors:

14.	15.	16.	17.
$? \times 9 = 108$	$? \times 2 = 56$	$? \times 11 = 132$	$8 \times ? = 168$
$7 \times ? = 98$	$3 \times ? = 63$	$6 \times ? = 102$	$? \times 12 = 120$

State quotients and remainders:

18.	$8\overline{)90}$	$12\overline{)65}$	$11\overline{)79}$	$7\overline{)38}$	$9\overline{)85}$	$8\overline{)75}$
19.	$7\overline{)45}$	$9\overline{)38}$	$8\overline{)43}$	$11\overline{)36}$	$12\overline{)50}$	$10\overline{)98}$
20.	$41 \div 5$	$19 \div 7$	$38 \div 4$	$37 \div 3$	$36 \div 8$	$34 \div 6$

We may write these examples also as fractions:

$$\frac{41}{5} \qquad \frac{19}{7}$$

The line between indicates division.

21. How many 8¢ notebooks can you buy for 90¢? How much change will you receive?

Divide:

22. $\frac{35}{8}$ $\frac{44}{7}$ $\frac{61}{12}$ $\frac{88}{10}$ $\frac{80}{11}$ $\frac{38}{9}$ $\frac{40}{7}$ $\frac{26}{5}$

23. $\frac{24}{5}$ $\frac{36}{7}$ $\frac{80}{11}$ $\frac{100}{8}$ $\frac{50}{6}$ $\frac{67}{12}$ $\frac{88}{10}$ $\frac{30}{8}$

24. $\frac{49}{8}$ $\frac{75}{12}$ $\frac{90}{11}$ $\frac{30}{7}$ $\frac{39}{5}$ $\frac{52}{7}$ $\frac{36}{10}$ $\frac{18}{4}$

WRITTEN PROBLEMS

1. How long will it take an automobile to travel 644 miles at the rate of 28 miles an hour?

2. If 21 horses cost $3024, how much will one horse cost?

3. If $3968 is divided equally among 62 people, how much will each receive?

4. How many automobile tires can be bought for $4068, if each one costs $36?

5. A company delivers 16 loads of coal making 63,184 pounds in all. Find the average size of each load.

WRITTEN EXERCISE

Divide:

1. $3036 \div 23$
2. $6888 \div 44$
3. $5275 \div 25$
4. $6783 \div 21$
5. $7310 \div 34$
6. $2806 \div 46$
7. $2016 \div 84$
8. $9787 \div 42$
9. $2828 \div 23$
10. $1835 \div 41$
11. $2709 \div 63$
12. $6572 \div 62$
13. $9745 \div 84$
14. $6572 \div 31$
15. $8904 \div 84$
16. $8910 \div 42$
17. $3294 \div 27$
18. $5692 \div 82$
19. $9162 \div 43$
20. $3816 \div 72$
21. $1188 \div 36$
22. $1458 \div 54$
23. $1404 \div 27$
24. $4672 \div 64$

WRITTEN EXERCISE

Divide, writing the remainders as fractions:

1. $8904 \div 43$
2. $5276 \div 25$
3. $9263 \div 43$
4. $7520 \div 25$
5. $6783 \div 22$
6. $7355 \div 34$
7. $5464 \div 43$
8. $6872 \div 52$
9. $1343 \div 61$
10. $1474 \div 72$
11. $1682 \div 81$
12. $1845 \div 91$
13. $2000 \div 19$
14. $5554 \div 43$
15. $1267 \div 32$
16. $3436 \div 26$
17. $1547 \div 25$
18. $2225 \div 32$
19. $1520 \div 37$
20. $1348 \div 42$
21. $1348 \div 34$
22. $7575 \div 71$
23. $2425 \div 26$
24. $8615 \div 37$

Find quotients and remainders:

25. $16{,}992 \div 72$
26. $466{,}344 \div 216$
27. $58{,}563 \div 9$
28. $70{,}449 \div 60$
29. $70{,}855 \div 34$
30. $342{,}055 \div 211$
31. $57{,}675 \div 43$
32. $586{,}197 \div 221$
33. $106{,}710 \div 31$
34. $260{,}068 \div 98$
35. $171{,}987 \div 130$
36. $450{,}039 \div 71$
37. $58{,}563 \div 81$
38. $80{,}960 \div 59$
39. $111{,}172 \div 21$
40. $300{,}176 \div 73$
41. $69{,}483 \div 48$
42. $233{,}159 \div 18$

ORAL EXERCISE

Supply the missing factors and multiples:

1.	2.	3	4.
?÷7=14	84÷?=4	?÷12=6	?÷5=17
?÷8=11	?÷8=15	132÷?=11	110÷?=11
117÷?=13	?÷26=2	?÷11=12	?÷7=12
?÷11=10	120÷?=10	?÷9=12	144÷?=12
112÷?=14	?÷17=4	91÷?=13	108÷?=18

Divide:

5.	80)960	14)\$4.20	90)8100	70)\$8.40
6.	11)2200	12)2400	50)\$7.50	18)\$3.60
7.	60)780	200)200	50)1500	60)720
8.	30)\$9.00	15)\$4.50	40)\$16.00	18)\$5.40
9.	90)4500	40)1200	50)\$30.00	17)\$5.10
10.	20)\$14.00	12)240	120)\$3.60	600)\$54.00

Dividends with Dollars and Cents.

Divide \$433.86 by 21:

```
      $20.66
21)$433.86
     42
     —
     13 8
     12 6
     ————
      1 26
      1 26
      ————
```

Notice that we place the decimal point in the quotient directly above the decimal point in the dividend.

The example means that if we divide \$433.86 into 21 equal parts, there will be \$20.66 in each part.

Check: \$20.66 × 21 = \$433.86.

WRITTEN EXERCISE

Divide:

1. \$8.68÷28	7. \$1344.78÷31	13. \$1944.72÷36
2. \$5.40÷12	8. \$4473.84÷21	14. \$5588.22÷22
3. \$74.48÷49	9. \$5313.69÷23	15. \$3936.94÷32
4. \$277.02÷19	10. \$300.76÷73	16. \$5566÷22
5. \$96.96÷16	11. \$840.99÷97	17. \$340,083÷31
6. \$169.92÷36	12. \$1844.16÷24	18. \$4416.92÷46

Dividing by Three Figures.

Divide 72,020 by 143:

$$\begin{array}{r} 503\frac{91}{143} \\ 143\overline{)72,020} \\ 71\,5 \\ \hline 520 \\ 429 \\ \hline 91 \end{array}$$

In dividing by three figures, we use the first two figures of the dividend as a trial divisor. $72 \div 14 = 5$, so that 720 contains 143 about 5 times.

When we subtract, we have a remainder of 5.

We bring down 2. This gives 52 as a partial dividend. 143 is not contained in 52 so we are careful to place a 0 above the 1 in the quotient.

Proof: $503 \times 143 + 91 = 72,020$.

WRITTEN EXERCISE

1. 1324÷331	6. 4598÷418	11. \$82,450÷388
2. 1561÷243	7. 6714÷746	12. 35,256÷113
3. 2752÷334	8. 5873÷839	13. 61,280÷152
4. 2947÷334	9. \$527.88÷212	14. 84,581÷281
5. 2484÷276	10. \$288.75÷231	15. \$1871.76÷264

Divide:

16. 70,060 ÷ 226	**20.** \$5763.32 ÷ 143	**24.** \$2401.60 ÷ 395
17. \$873.54 ÷ 211	**21.** \$2899.31 ÷ 151	**25.** \$3298.32 ÷ 342
18. 85,407 ÷ 249	**22.** 186,912 ÷ 132	**26.** \$6455.82 ÷ 399
19. 92,008 ÷ 217	**23.** 561,446 ÷ 347	**27.** \$7521.06 ÷ 618

Dividing by Multiples of 10.

1. 10)30 10)300 1000)3000 400)1200

1Ø)3Ø 3	1Ø)3ØØ 3	1ØØØ)3ØØØ 3	4ØØ)12ØØ 3

2. 400)3200 400)3217 4000)32,172

4ØØ)32ØØ 8	4ØØ)3217 $8\frac{17}{400}$	4ØØØ)32,172 $8\frac{172}{4000}$

Cancel the zeros at the right of the divisor. Cancel as many figures at the right of the dividend as you cancel zeros in the divisor. Write the complete remainder over the complete divisor.

WRITTEN EXERCISE

Divide:

1. 540 ÷ 90	**6.** 16,200 ÷ 300	**11.** 34,000 ÷ 2000
2. 840 ÷ 70	**7.** 33,504 ÷ 500	**12.** 42,009 ÷ 2000
3. 5600 ÷ 80	**8.** 51,112 ÷ 700	**13.** 114,000 ÷ 3000
4. 8402 ÷ 200	**9.** 50,400 ÷ 1200	**14.** 152,027 ÷ 4000
5. 5504 ÷ 500	**10.** 64,019 ÷ 1400	**15.** 520,116 ÷ 4000

WRITTEN REVIEW PROBLEMS

1. How many pieces of rope each 21 feet in length can be cut from a coil of 1554 feet?

2. 73 people have a total amount of $30,076 deposited in a bank. Find what the average sum is for each one.

3. A machine made 16,992 nails, which are to be packed in boxes each holding 6 doz. How many boxes will be needed?

4. A dealer paid $2952 for a shipment of 41 typewriters. He sold them all for $3485. How much did he gain on each machine?

5. In a school containing 2193 pupils, there are 51 classrooms. What is the average number of pupils in each room?

6. A dealer sold 38 large chairs for $798. How much did he receive for each chair?

7. In 3597 quarts of barley how many pecks will there be? How many quarts will be left over?

8. How long would a train require to travel 1950 miles at the rate of 30 miles an hour?

9. If an acre will produce 16 bushels of wheat, how many acres will be required to produce 10,944 bushels?

10. A party of 15 boys went camping. The fares were $67.40, the food cost $122, and other expenses $76.10. What was each boy's share of the expenses?

11. A dealer bought 3 dozen umbrellas at $.85 each. He sold $\frac{3}{4}$ of them at $1.25 and the remainder at $.98. Find his gain.

ORAL DRILL EXERCISE

	A	*B*	*C*	*D*
1.	63÷9	81÷9	\$5−\$1.25	48÷6
2.	10×47	\$2−\$.49	48÷8	36¢×10
3.	56÷8	64÷8	54¢×10	45÷5
4.	39¢×20	10×69	\$3−\$.39	10×73
5.	49÷5	47÷8	80÷9	52÷9
6.	3 yr. = ? mo.	$\frac{1}{2}$ doz. = ?	8 pt. = ? qt.	2 T. = ? lb.
7.	$\frac{1}{2}$ bu. = ? pk.	5 wk. = ? da.	1 yr. = ? wk.	2 yd. = ? in.
8.	16 gal. = ? qt.	$\frac{3}{4}$ pk. = ? qt.	15 ft. = ? yd.	48 oz. = ? lb.
9.	6 ft. = ? in.	4 hr. = ? min.	$\frac{1}{2}$ yd. = ? ft.	50 in. = ? ft.

Give answers:

10.	19¢×20	96÷12	14×7	36¢×20
11.	8×15	9÷40	6×35	11×11
12.	10)49	4×21	10)95	11)80
13.	87¢−58¢	92¢−39¢	46¢−19¢	75¢−48¢
14.	27¢+44¢	$\frac{1}{5}$ of 90	46¢+77¢	$\frac{7}{10}$ of 40
15.	$\frac{1}{2}$ of 84	25¢+77¢	$\frac{3}{4}$ of 44	55¢+24¢
16.	54¢+97¢	$\frac{2}{3}$ of 39	11×12	4×17
17.	3×24	5×19	58¢+22¢	$\frac{4}{5}$ of 65
18.	$\frac{3}{8}$ of 32	88¢+33¢	$\frac{1}{8}$ of 72	78¢+77¢
19.	$14\frac{1}{2}$ $+1\frac{1}{4}$	$15\frac{3}{4}$ $+2\frac{1}{4}$	$16\frac{1}{4}$ $+3\frac{3}{4}$	$26\frac{3}{4}$ $+2\frac{1}{4}$
20.	$2\frac{1}{4}$ $5\frac{1}{4}$ $1\frac{1}{4}$	$3\frac{1}{4}$ 4 $2\frac{3}{4}$	$6\frac{1}{4}$ $1\frac{1}{2}$ $3\frac{1}{4}$	$5\frac{1}{4}$ $4\frac{1}{4}$ $2\frac{1}{2}$

GENERAL ORAL PROBLEMS

1. A car uses $3\frac{1}{3}$ gal. of oil on each trip it makes. How many gal. will it use on 3 trips?

2. Helen bought 4 blank books at 9¢ each. How many 3¢ lead pencils could she have bought with the same money?

3. How much will 5 cans of pepper cost at 36¢ a can?

4. If 2 lb. of coffee cost 60¢, what will $1\frac{1}{2}$ lb. cost?

5. Harry bought an air rifle for $1.85 and sold it later for $.70 less. How much did he get for it?

6. Ethel spent $\$\frac{1}{5}$ for candy and $\$\frac{1}{10}$ for carfare. How many cents did she spend?

7. If 5 eggs cost 20¢, how much will 1 doz. cost?

8. Find the cost of $2\frac{1}{4}$ lb. of rice at 8¢ a pound.

9. It took mother 34 min. to go downtown and 56 min. to come back. How long was she on the car?

10. How many pecks in 112 quarts?

11. If shirts are 3 for a dollar, what will 18 cost?

12. Find the cost of a dozen tennis rackets at $1.50 each.

13. A messenger boy left a store at 10:40 A.M. and returned at 1 P.M. How long was he away?

14. What fraction of a pound is two ounces?

15. How many days in September, October, and November? In May, June, and July?

16. If I buy 4 qt. of potatoes at 12¢ and a can of peaches for 21¢, how much change shall I have from a dollar bill?

GENERAL WRITTEN PROBLEMS

1. An agent bought a house for $6250 and a lot for $4325. Find how much he must sell them for to gain $2832.

2. We pay $33 a month for rent, and $3.00 a month for gas. How much do we pay for rent and gas a year?

3. The grocer bought 17 bushels of potatoes at $.55 a bushel, and 2 barrels of cranberries at $6.75 a barrel. How much did he pay for all?

4. How many seats are there in 18 rooms each containing 42 seats?

5. 8 cows were sold at $62 each and 3 horses at $135 each. Find the amount of the bill.

6. How many bushels are there in 336 packs?

7. An employer divided $69.66 among 18 helpers. How much did each receive?

8. A man had $294 in the bank. How much had he there after drawing out $\frac{4}{7}$ of it?

9. Find the cost of 3 doz. armchairs at $22.60 and 14 pictures at $4.75 each.

10. If $\frac{1}{2}$ a yd. of carpet costs $3.50, what will 20 yd. cost?

11. A farmer raised 714 bu. of potatoes and sold $\frac{5}{6}$ of them at 70¢ a peck. How much did he receive?

12. A factory paid out $18,375.12 in wages in one week. Find the average amount earned each working day.

VII. FRACTIONS

ORAL EXERCISE

1. How do you find $\frac{1}{2}$ of a line? $\frac{1}{4}$? $\frac{1}{8}$?

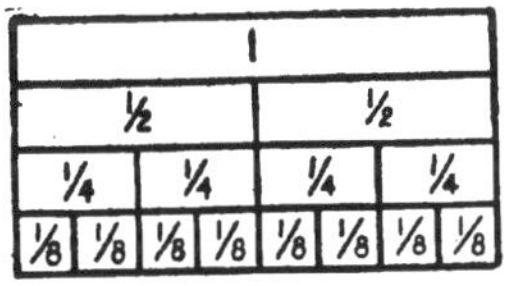

2. How many quarters in one line? How many quarters in one half a line? How many eighths in $\frac{1}{2}$? In $\frac{1}{4}$?

3. How do you find $\frac{3}{4}$ of a line? What does the four tell us about the line? What does the three tell us? How many fourths in $\frac{3}{4}$? How many eighths in $\frac{1}{2}$? In $\frac{1}{4}$?

4. How many halves in 1? In 2? In 3? In $1\frac{1}{2}$? In $2\frac{1}{2}$? How many whole numbers are there in $\frac{2}{2}$? In $\frac{4}{2}$? In $\frac{8}{2}$? In $\frac{3}{2}$? In $\frac{7}{2}$? In $2\frac{1}{4}$?

5. How many fourths in $1\frac{3}{4}$? How many in 2?

6. How many eighths in $1\frac{1}{8}$? In $1\frac{1}{4}$? In $1\frac{1}{2}$? In 2?

7. How many whole numbers in $\frac{5}{4}$? In $\frac{8}{8}$? In $\frac{9}{4}$? In $\frac{10}{8}$? In $\frac{24}{8}$?

8. Draw a figure as above. Divide it into 6 equal parts. What is each part called? What would you call 2 parts of this line? How would you find $\frac{5}{6}$ of this line?

9. Draw a figure as above. Divide it into 9 equal parts. What is each part called? What would you call 3 parts of this line? 6 parts? How would you find $\frac{2}{9}$ of the line? $\frac{5}{9}$? $\frac{7}{9}$? $\frac{8}{9}$?

ORAL EXERCISE

1. Read the following and tell which figure shows the number of equal parts into which a unit has been divided and which shows how many of these equal parts have been taken:

$\frac{1}{3}$ $\frac{2}{3}$ $\frac{5}{6}$ $\frac{3}{8}$ $\frac{7}{8}$ $\frac{1}{9}$ $\frac{4}{9}$ $\frac{1}{5}$ $\frac{4}{5}$ $\frac{1}{12}$ $\frac{7}{12}$ $\frac{1}{11}$

Find:

2.	3.	4.	5.	6.
$\frac{1}{2}$ of 4	$\frac{1}{5}$ of 20	$\frac{1}{3}$ of 9	$\frac{3}{4}$ of 16	$\frac{1}{6}$ of 12
$\frac{1}{4}$ of 4	$\frac{1}{2}$ of 14	$\frac{1}{3}$ of 15	$\frac{4}{5}$ of 10	$\frac{5}{6}$ of 6
$\frac{1}{3}$ of 6	$\frac{1}{8}$ of 24	$\frac{2}{3}$ of 9	$\frac{1}{2}$ of 20	$\frac{3}{8}$ of 16
$\frac{1}{3}$ of 9	$\frac{1}{3}$ of 30	$\frac{3}{3}$ of 9	$\frac{1}{5}$ of 20	$\frac{1}{2}$ of 24

7. How many days in a week? What part of a week is 1 day? 4 days? 6 days? 5 days? 3 days?

8. How many quarts in a peck? What part of a peck is 1 quart? 3 quarts? 7 quarts?

9. How many months in a year? What part of a year is 1 month? 6 months? 7 months? 5 months? 11 months?

WRITTEN EXERCISE

Find:

1.	2.	3.	4.	5.
$\frac{2}{3}$ of 36	$\frac{2}{7}$ of 84	$\frac{3}{5}$ of 85	$\frac{5}{8}$ of 96	$\frac{5}{8}$ of 72
$\frac{7}{8}$ of 48	$\frac{4}{10}$ of 90	$\frac{5}{11}$ of 77	$\frac{7}{12}$ of 84	$\frac{8}{9}$ of 81
$\frac{5}{6}$ of 72	$\frac{6}{7}$ of 49	$\frac{3}{4}$ of 44	$\frac{4}{5}$ of 55	$\frac{2}{7}$ of 84
$\frac{3}{8}$ of 96	$\frac{5}{8}$ of 64	$\frac{3}{7}$ of 35	$\frac{7}{8}$ of 96	$\frac{4}{9}$ of 45
$\frac{1}{8}$ of 88	$\frac{1}{9}$ of 81	$\frac{1}{11}$ of 66	$\frac{1}{12}$ of 48	$\frac{3}{5}$ of 75

Terms in Fractions. A whole number, such as 4, 8, 12, 24, is called an **integer.**

A **fraction** means that a part of something has been taken.

$\frac{3}{4}$ The **numerator** is the number above the line.
The **denominator** is the number below the line.

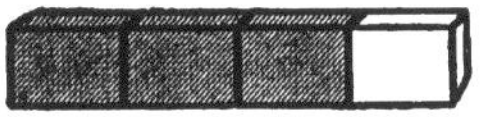

We have taken $\frac{3}{4}$ of this block. The *denominator* 4 tells the number of equal parts into which the block has been divided. The *numerator* 3 tells how many of these parts we have taken.

The numerator and the denominator, 3 and 4 in this case, are called the **terms** of the fraction.

A number consisting of an integer and a fraction written together is called a **mixed number.** $2\frac{1}{4}$, $3\frac{1}{2}$ are mixed numbers.

A fraction that shows a part of a unit, such as $\frac{1}{2}$, $\frac{3}{4}$, $\frac{2}{3}$, is a **proper fraction.** Proper fractions are *less* than 1.

A fraction that is *equal* to 1 unit, like $\frac{2}{2}$, $\frac{4}{4}$, $\frac{3}{3}$, or is *greater* than a unit, like $\frac{3}{2}$, $\frac{5}{4}$, $\frac{4}{3}$, is called an **improper fraction.**

We always try to change improper fractions to whole numbers or to mixed numbers.

$\frac{3}{3}=1$ $\frac{4}{3}=1\frac{1}{3}$ $\frac{8}{8}=1$ $\frac{9}{8}=1\frac{1}{8}$ $\frac{11}{8}=1\frac{3}{8}$ $\frac{7}{6}=1\frac{1}{6}$

ORAL EXERCISE

1. Tell which are integers and which are mixed numbers:

$8 \quad 2\frac{3}{4} \quad 1\frac{1}{4} \quad 9 \quad 5 \quad 5\frac{4}{5} \quad 9\frac{2}{3} \quad 7 \quad 1\frac{7}{8}$

2. Tell which are proper fractions and which are improper fractions:

$\frac{2}{2} \quad \frac{7}{8} \quad \frac{4}{4} \quad \frac{5}{2} \quad \frac{6}{4} \quad \frac{7}{7} \quad \frac{5}{6} \quad \frac{11}{3} \quad \frac{9}{9} \quad \frac{6}{4}$

Find:

3.	4.	5.	6.
$\frac{1}{3}$ of 39	$\frac{2}{5}$ of 70	$\frac{5}{6}$ of 60	$\frac{6}{7}$ of 35
$\frac{3}{8}$ of 88	$\frac{2}{3}$ of 300	$\frac{3}{4}$ of 60	$\frac{4}{5}$ of 60
$\frac{3}{10}$ of 90	$\frac{4}{11}$ of 99	$\frac{10}{11}$ of 77	$\frac{3}{4}$ of 100
$\frac{5}{12}$ of 72	$\frac{2}{9}$ of 54	$\frac{7}{8}$ of 104	$\frac{1}{9}$ of 99
$\frac{1}{16}$ of 48	$\frac{5}{7}$ of 84	$\frac{5}{6}$ of 300	$\frac{3}{16}$ of 32

ORAL PROBLEMS

1. Ethel spent 20¢ for a handkerchief and $\frac{3}{4}$ as much for a ribbon. How much did she spend for both?

2. Harry wrote 42 spelling words and $\frac{1}{7}$ of them were wrong. How many were correct?

3. Butter is 36¢ a lb. How much did Gertrude pay for $\frac{3}{4}$ of a lb.? How much did $1\frac{1}{4}$ lb. cost?

4. $\frac{1}{5}$ of a yd. of silk costs 25¢. Find the cost of 1 yd.

5. Margaret had 40¢ and she spent $\frac{3}{5}$ of it in the dry-goods store. How much did she spend?

6. A boy had 24 marbles. How many did he lose if he lost $\frac{1}{2}$ of them? $\frac{1}{3}$? $\frac{1}{4}$? $\frac{3}{4}$? $\frac{1}{6}$? $\frac{5}{6}$? $\frac{1}{8}$? $\frac{3}{8}$? $\frac{7}{8}$? $\frac{1}{12}$? $\frac{5}{12}$? $\frac{7}{12}$?

WRITTEN PROBLEMS

1. Joe's bicycle cost \$27. Ted bought a good one for $\frac{7}{9}$ of the price of Joe's. How much did Ted's cost?

2. Harry's father gave him \$35 to pay his expenses on a camping trip. He spent only $\frac{6}{7}$ of the money. How much did he spend?

3. A horse cost \$152. A cow was bought for $\frac{3}{8}$ of that amount. How much did the cow cost?

4. Mary had 75¢ in the school bank. She drew out $\frac{3}{5}$ of the amount. How much did she leave in?

5. We had 228 plants in the school garden. A storm blew down $\frac{1}{6}$ of them. How many were blown down?

WRITTEN EXERCISE

Find values:

1. $\frac{1}{3}$ of 144
2. $\frac{1}{3}$ of 264
3. $\frac{2}{3}$ of 252
4. $\frac{1}{4}$ of 288
5. $\frac{3}{4}$ of 320
6. $\frac{1}{3}$ of 450
7. $\frac{2}{3}$ of 510
8. $\frac{3}{4}$ of 524
9. $\frac{1}{5}$ of 330
10. $\frac{3}{5}$ of 675
11. $\frac{4}{5}$ of 725
12. $\frac{1}{6}$ of 336
13. $\frac{5}{6}$ of 726
14. $\frac{2}{5}$ of 770
15. $\frac{1}{6}$ of 378
16. $\frac{5}{6}$ of 756
17. $\frac{1}{7}$ of 847
18. $\frac{3}{7}$ of 588
19. $\frac{5}{7}$ of 287
20. $\frac{1}{8}$ of 448
21. $\frac{7}{8}$ of 512
22. $\frac{3}{7}$ of 287
23. $\frac{4}{7}$ of 294
24. $\frac{3}{8}$ of 488
25. $\frac{5}{8}$ of 488
26. $\frac{7}{8}$ of 328
27. $\frac{1}{9}$ of 729
28. $\frac{5}{9}$ of 720
29. $\frac{6}{9}$ of 639
30. $\frac{7}{8}$ of 499
31. $\frac{2}{9}$ of 558
32. $\frac{7}{9}$ of 738
33. $\frac{4}{9}$ of 657

Multiplying by a Mixed Number.

1. A yard of silk cost 96¢. How much will $4\frac{1}{4}$ yards cost?

2. An automobile goes 24 miles an hour. How far will it travel in $7\frac{2}{3}$ hours?

\$.96	24
$4\frac{1}{4}$	$7\frac{2}{3}$
24 = $\frac{1}{4}$ of 96	16 = $\frac{2}{3}$ of 24
384	168
\$4.08	184 miles

3. A girl bought $8\frac{1}{4}$ yd. of cloth at \$.48 a yd. How much change should she get from \$5?

4. A grocer buys grape jelly in jars at \$2.70 a doz. How much will he pay for an order of $9\frac{2}{3}$ doz. jars?

5. If apples sell at 30¢ a peck, how much will $3\frac{1}{2}$ bushels cost?

6. If shelled walnuts sell for 56¢ a pound, how much shall I have to pay for $6\frac{3}{4}$ lb.?

7. Find the cost of $9\frac{1}{8}$ lb. tea at 48¢ a lb.

WRITTEN EXERCISE

Find the products:

1. $84\times2\frac{1}{4}$	5. $55\times5\frac{2}{5}$	9. $33\times6\frac{2}{3}$	13. $42\times10\frac{1}{7}$
2. $70\times5\frac{1}{5}$	6. $44\times4\frac{3}{4}$	10. $65\times8\frac{1}{5}$	14. $55\times12\frac{1}{11}$
3. $88\times4\frac{1}{2}$	7. $96\times4\frac{1}{4}$	11. $30\times3\frac{3}{5}$	15. $64\times8\frac{3}{8}$
4. $54\times2\frac{1}{8}$	8. $60\times4\frac{5}{6}$	12. $56\times7\frac{1}{8}$	16. $36\times12\frac{3}{4}$

ORAL EXERCISE

1. To find $\frac{1}{5}$ of a ruler, into how many equal parts must the ruler be divided? How would you find $\frac{1}{6}$ of a number? $\frac{3}{8}$ of a pie? $\frac{6}{7}$ of a foot? $\frac{2}{9}$ of a gallon? $\frac{7}{12}$ of a pound?

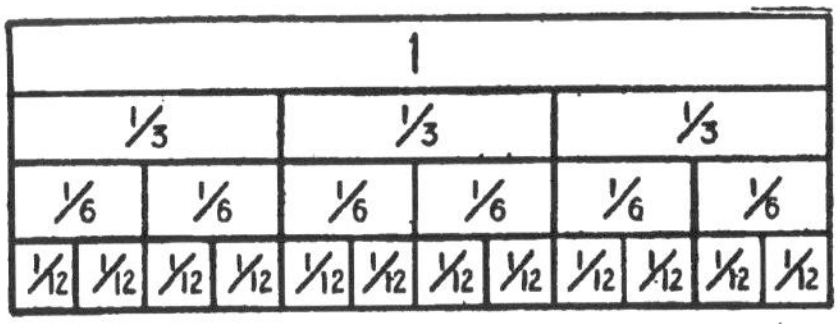

2. How many sixths in $\frac{1}{3}$? How many sixths in $\frac{2}{3}$? In $\frac{1}{2}$? $\frac{2}{2}=\frac{?}{6}$ $\frac{2}{6}=\frac{?}{3}$

3. $\frac{1}{6}+\frac{1}{3}=\frac{?}{6}$ $\frac{1}{2}+\frac{1}{3}=\frac{?}{6}$ $\frac{4}{6}=\frac{?}{3}$ $\frac{6}{6}=\frac{?}{2}$ $\frac{2}{3}+\frac{1}{6}=\frac{?}{6}$ $\frac{1}{3}-\frac{1}{6}=\frac{?}{6}$ $\frac{2}{3}-\frac{1}{6}=?$ $\frac{5}{6}-\frac{1}{3}=?$ $\frac{4}{6}-\frac{1}{3}=?$ $\frac{4}{6}-\frac{2}{3}=?$

4. How many twelfths in $\frac{1}{6}$? $\frac{1}{3}$? In $\frac{2}{3}$? In $\frac{5}{6}$? In $\frac{3}{4}$? $\frac{4}{6}=\frac{?}{12}$ $\frac{2}{12}=?$ $\frac{4}{12}=?$ $\frac{6}{12}=?$ $\frac{8}{12}=?$ $\frac{10}{12}=?$

5. $\frac{1}{6}+\frac{1}{6}=?$ $\frac{1}{12}+\frac{1}{12}=?$ $\frac{5}{12}+\frac{1}{12}=?$ $\frac{11}{12}+\frac{1}{12}=?$ $\frac{1}{3}+\frac{1}{12}=?$ $\frac{7}{12}-\frac{1}{12}=?$ $\frac{1}{3}-\frac{1}{12}=?$ $1-\frac{5}{12}=?$ $\frac{1}{6}+\frac{1}{12}=?$ $\frac{8}{12}-\frac{2}{6}=?$

ORAL EXERCISE

1. How many fifths in 1? In 3? In $1\frac{2}{5}$? In $4\frac{4}{5}$?

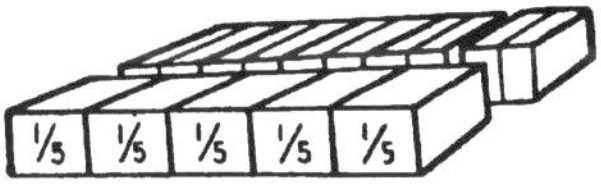

2. How many tenths in $\frac{1}{5}$? In $\frac{2}{5}$? $\frac{10}{10}=\frac{?}{5}$ $\frac{4}{10}=\frac{?}{5}$ $\frac{1}{10}+\frac{1}{5}=\frac{?}{10}$ $\frac{1}{5}-\frac{1}{10}=\frac{?}{10}$ $4\frac{3}{10}+4\frac{3}{10}=?$ $7\frac{5}{10}+1\frac{3}{10}=?$

3. How many tenths in 2? In 3? In $2\frac{3}{10}$? In $4\frac{1}{2}$?

4. How many whole numbers in $\frac{20}{5}$? $\frac{12}{5}$? $\frac{11}{10}$? $\frac{19}{10}$?

5. There are 75 people in a car. $\frac{3}{5}$ of them asked for transfers. How many was that?

6. $\frac{3}{10}$ of 40 is how many? $\frac{7}{10}$ of 80?

Reduction of Fractions to Lower and Higher Terms.

Reduce $\frac{1}{2}$ to higher terms: $=\frac{1\times 2}{2\times 2}=\frac{2}{4}$

1. How many fourths in $\frac{1}{2}$? How many sixths? How many eighths? Tenths? Twelfths?

2. How many thirds in $\frac{4}{12}$? $\frac{4}{12}=\frac{4\div 4}{12\div 4}=\frac{1}{3}$ How many thirds in $\frac{3}{9}$? In $\frac{2}{6}$?

3. What is the effect on the value of a fraction, of multiplying both terms by the same number? What is the effect of dividing both terms by the same number?

Both terms of a fraction may be multiplied by the same number or divided by the same number without changing the value of the fraction.

This process of changing the terms of a fraction without changing its value is called **reduction.**

4. When the terms of a fraction are changed to smaller numbers without changing the value of the fraction, the fraction is reduced to *lower terms.* Which has the lower terms: $\frac{6}{12}$ or $\frac{1}{2}$?

5. When the terms of a fraction are changed to larger numbers, the fraction is raised to *higher terms.* Which has the higher terms: $\frac{4}{16}$ or $\frac{1}{4}$?

$\frac{1}{2}$, $\frac{4}{5}$, $\frac{2}{7}$, $\frac{5}{16}$ cannot be changed to lower terms.

A fraction is in its *lowest terms* when its terms can be divided by no number greater than one.

To reduce a fraction to lower terms, divide both its terms by the same number.

To raise a fraction to higher terms, multiply both its terms by the same number.

ORAL EXERCISE

Tell which of the following fractions are in their lowest terms. Reduce the others to lowest terms:

1. $\frac{1}{2}$ $\quad \frac{4}{16}$ $\quad \frac{6}{8}$ $\quad \frac{2}{4}$ $\quad \frac{4}{6}$ $\quad \frac{8}{12}$ $\quad \frac{3}{7}$ $\quad \frac{7}{9}$
2. $\frac{8}{10}$ $\quad \frac{3}{12}$ $\quad \frac{2}{9}$ $\quad \frac{4}{8}$ $\quad \frac{4}{11}$ $\quad \frac{4}{10}$ $\quad \frac{5}{20}$ $\quad \frac{10}{15}$
3. $\frac{5}{16}$ $\quad \frac{5}{12}$ $\quad \frac{4}{12}$ $\quad \frac{10}{14}$ $\quad \frac{5}{6}$ $\quad \frac{3}{6}$ $\quad \frac{9}{9}$ $\quad \frac{5}{8}$
4. $\frac{5}{15}$ $\quad \frac{15}{20}$ $\quad \frac{3}{9}$ $\quad \frac{6}{9}$ $\quad \frac{2}{10}$ $\quad \frac{3}{16}$ $\quad \frac{3}{8}$ $\quad \frac{4}{20}$

Reduce to higher terms:

5.	6.	7.	8.	9.
$\frac{1}{2}=\frac{?}{20}$	$\frac{2}{7}=\frac{?}{14}$	$\frac{2}{9}=\frac{?}{18}$	$\frac{3}{4}=\frac{?}{24}$	$\frac{1}{6}=\frac{?}{18}$
$\frac{1}{3}=\frac{?}{12}$	$\frac{1}{8}=\frac{?}{16}$	$\frac{1}{2}=\frac{?}{18}$	$\frac{3}{5}=\frac{?}{20}$	$\frac{5}{6}=\frac{?}{18}$
$\frac{5}{6}=\frac{?}{12}$	$\frac{5}{8}=\frac{?}{24}$	$\frac{2}{3}=\frac{?}{9}$	$\frac{2}{5}=\frac{?}{15}$	$\frac{2}{3}=\frac{?}{15}$

WRITTEN EXERCISE

1. Reduce to halves:

   $1 \quad \frac{2}{4} \quad \frac{8}{4} \quad 2 \quad \frac{8}{8} \quad 3 \quad \frac{10}{10} \quad \frac{6}{12} \quad \frac{32}{8}$

2. Reduce to fourths:

   $\frac{1}{2} \quad \frac{2}{8} \quad \frac{4}{8} \quad 1 \quad \frac{10}{8} \quad \frac{8}{16} \quad 3 \quad \frac{2}{2} \quad 4$

3. Reduce to fifths:

   $\frac{3}{15} \quad \frac{10}{10} \quad 2 \quad \frac{30}{10} \quad 3 \quad \frac{5}{20} \quad \frac{2}{10} \quad \frac{6}{30} \quad \frac{10}{25}$

4. Reduce to eighths:

$\frac{1}{2}$ $\frac{3}{2}$ $\frac{1}{4}$ $\frac{3}{4}$ 1 $\frac{5}{4}$ 2 $\frac{8}{16}$ $\frac{4}{4}$

5. Reduce to sixths:

$\frac{1}{2}$ $\frac{1}{3}$ $\frac{2}{3}$ $\frac{6}{12}$ $\frac{9}{18}$ 3 $\frac{2}{2}$ $\frac{3}{3}$ $\frac{4}{2}$

6. Reduce to twelfths:

$\frac{1}{2}$ 2 $\frac{1}{3}$ $\frac{2}{3}$ $\frac{3}{4}$ $\frac{1}{6}$ $\frac{5}{6}$ $\frac{4}{24}$ 3

7. Reduce to lowest terms:

$\frac{12}{14}$ $\frac{3}{18}$ $\frac{8}{14}$ $\frac{6}{16}$ $\frac{12}{18}$ $\frac{4}{30}$ $\frac{2}{40}$ $\frac{7}{7}$ $\frac{16}{20}$

Reduction of Improper Fractions to Mixed Numbers.

$\frac{1}{2}$ lb. + $\frac{1}{2}$ lb. + $\frac{1}{2}$ lb. = $\frac{?}{2}$ lb.
$\frac{1}{3}$ ft. + $\frac{1}{3}$ ft. + $\frac{1}{3}$ ft. + $\frac{1}{3}$ ft. = $\frac{?}{3}$ ft.
$\frac{3}{4}$ hr. + $\frac{3}{4}$ hr. + $\frac{3}{4}$ hr. = $\frac{?}{4}$ hr.

The answers to these three examples are **improper fractions.** We always change an improper fraction to a whole number or to a mixed number.

$\frac{3}{2}$ lb. = $1\frac{1}{2}$ lb. $\frac{4}{3}$ ft. = $1\frac{1}{3}$ ft. $\frac{9}{4} = 9 \div 4 = 2\frac{1}{4}$, so that $\frac{9}{4}$ hr. = $2\frac{1}{4}$ hr.

To reduce an improper fraction to a whole number or to a mixed number, divide the numerator by the denominator.

WRITTEN EXERCISE

Reduce to whole or mixed numbers:

1. $\frac{6}{2}$; $\frac{5}{2}$; $\frac{4}{3}$.
2. $\frac{7}{2}$; $\frac{9}{3}$; $\frac{8}{4}$.
3. $\frac{6}{3}$; $\frac{13}{3}$; $\frac{16}{4}$.
4. $\frac{9}{4}$; $\frac{14}{4}$; $\frac{9}{6}$.
5. $\frac{5}{5}$; $\frac{9}{5}$; $\frac{13}{2}$.
6. $\frac{22}{6}$; $\frac{12}{7}$; $\frac{26}{5}$.
7. $\frac{41}{4}$; $\frac{30}{6}$; $\frac{20}{7}$.
8. $\frac{16}{8}$; $\frac{20}{8}$; $\frac{28}{7}$.
9. $\frac{19}{8}$; $\frac{18}{9}$; $\frac{21}{9}$.
10. $\frac{20}{10}$; $\frac{11}{9}$; $\frac{24}{10}$.
11. $\frac{20}{12}$; $\frac{39}{6}$; $\frac{60}{12}$.
12. $\frac{50}{8}$; $\frac{40}{9}$; $\frac{30}{7}$.

Reduction of Mixed Numbers to Improper Fractions.

How many quarters of an inch are there in $3\frac{3}{4}$ inches? In 1 inch there are $\frac{4}{4}$. In $3\frac{3}{4}$ inches there are $3\times\frac{4}{4}+\frac{3}{4}$. $3\times\frac{4}{4}=\frac{12}{4}$. $\frac{12}{4}+\frac{3}{4}=\frac{15}{4}$ in. This is the same as $3\frac{3}{4}=\frac{15}{4}$.

To reduce a mixed number to an improper fraction, multiply the integer by the denominator of the fraction, add the numerator, and place the result over the denominator.

WRITTEN EXERCISE

Reduce to improper fractions:

1. $7\frac{1}{2}$; $1\frac{1}{3}$; $3\frac{2}{3}$.	4. $2\frac{1}{5}$; $5\frac{3}{5}$; $4\frac{3}{4}$.	7. $4\frac{2}{7}$; $2\frac{1}{8}$; $9\frac{1}{8}$.
2. $3\frac{2}{5}$; $1\frac{1}{5}$; $6\frac{1}{4}$.	5. $6\frac{4}{5}$; $2\frac{1}{6}$; $7\frac{1}{5}$.	8. $3\frac{3}{8}$; $6\frac{5}{8}$; $7\frac{3}{8}$.
3. $4\frac{1}{4}$; $1\frac{3}{4}$; $2\frac{2}{3}$.	6. $2\frac{5}{6}$; $3\frac{1}{7}$; $8\frac{3}{4}$.	9. $2\frac{2}{9}$; $4\frac{3}{9}$; $2\frac{1}{12}$.

Use of Cancellation.

1. If 6 machines can turn out 84 hats, how many hats would be turned out by 24 machines?

Here we may divide 84 by 6 and find how many hats can be made by 1 machine. Then we may multiply this number by 24. Since we have learned, however, that the fraction $\frac{9}{4}$ means $9\div4$, we may indicate the division of 84 by 6 in this way: $\frac{84}{6}$. Then we may multiply that by 24, and shorten the work by striking out or canceling equal numbers.

$$\frac{84\times\overset{4}{\not{24}}}{\not{6}}=336 \text{ hats.}$$

2. When a field of 15 acres gives 220 bushels of wheat, how many bushels would 45 acres give?

WRITTEN EXERCISE

Find the value of the following:

1. $\frac{12\times3}{4}$	5. $\frac{7\times42}{6}$	9. $\frac{21\times6}{5\times7}$	13. $\frac{24\times44}{8\times11}$
2. $\frac{24\times5}{8}$	6. $\frac{33\times7}{16}$	10. $\frac{33\times7}{6\times11}$	14. $\frac{30\times14}{7\times10}$
3. $\frac{60\times7}{12}$	7. $\frac{3\times28}{7}$	11. $\frac{36\times15}{4\times12}$	15. $\frac{64\times72}{9\times8}$
4. $\frac{36\times11}{9}$	8. $\frac{16\times4}{3\times4}$	12. $\frac{45\times5}{3\times15}$	16. $\frac{48\times7}{4\times12}$

WRITTEN REVIEW EXERCISE

1. Reduce to lowest terms: $\frac{6}{8}$; $\frac{5}{10}$; $\frac{8}{12}$; $\frac{10}{15}$; $\frac{3}{18}$; $\frac{18}{20}$; $\frac{14}{16}$; $\frac{8}{30}$; $\frac{25}{30}$; $\frac{14}{21}$; $\frac{18}{24}$; $\frac{20}{32}$; $\frac{32}{36}$; $\frac{25}{45}$; $\frac{40}{48}$.

2. Reduce to whole or mixed numbers: $\frac{15}{2}$; $\frac{30}{4}$; $\frac{36}{5}$; $\frac{25}{8}$; $\frac{29}{7}$; $\frac{7}{5}$; $\frac{40}{9}$; $\frac{23}{6}$; $\frac{34}{10}$; $\frac{40}{11}$; $\frac{55}{17}$; $\frac{23}{14}$; $\frac{100}{16}$; $\frac{47}{15}$; $\frac{71}{12}$.

3. Reduce to improper fractions: $3\frac{3}{4}$; $4\frac{3}{7}$; $6\frac{5}{8}$; $5\frac{7}{8}$; $7\frac{2}{9}$; $8\frac{5}{9}$; $7\frac{2}{11}$; $8\frac{4}{5}$; $7\frac{8}{10}$; $6\frac{6}{7}$; $11\frac{1}{2}$; $6\frac{5}{12}$; $4\frac{7}{8}$.

Addition and Subtraction of Similar Fractions.

Add $4\frac{3}{5}+1\frac{2}{5}+\frac{1}{5}$:

$$\begin{array}{r} 4\frac{3}{5} \\ 1\frac{2}{5} \\ \frac{1}{5} \\ \hline \frac{6}{5}=1\frac{1}{5} \\ 5+1\frac{1}{5}=6\frac{1}{5} \end{array}$$

Fractions, such as the above, that have the same denominators are said to have a **common denominator** and are called **similar fractions.** Here we add the numerators, place the result over the denominator, and reduce if necessary.

ORAL EXERCISE

Add:

1.	$\frac{1}{3}$	$4\frac{1}{2}$	$1\frac{1}{3}$	$1\frac{1}{4}$	$5\frac{1}{4}$	$6\frac{2}{3}$	$7\frac{1}{2}$	$1\frac{2}{3}$
	$\frac{1}{3}$	$\frac{1}{2}$	$2\frac{2}{3}$	$\frac{2}{4}$	$\frac{1}{4}$	$1\frac{2}{3}$	$3\frac{1}{4}$	$\frac{2}{3}$
	$\frac{1}{3}$	$1\frac{1}{2}$	$3\frac{1}{3}$	$3\frac{1}{4}$	$1\frac{3}{4}$	1	$4\frac{1}{4}$	$1\frac{2}{3}$

Subtract:

2.	$4\frac{2}{3}$	$7\frac{3}{4}$	$7\frac{2}{3}$	$3\frac{2}{3}$	$9\frac{1}{2}$	$8\frac{1}{2}$	$7\frac{3}{4}$	$11\frac{3}{4}$
	$\frac{2}{3}$	$4\frac{2}{4}$	$5\frac{1}{3}$	$1\frac{2}{3}$	$1\frac{1}{2}$	$7\frac{1}{2}$	$7\frac{1}{4}$	$9\frac{3}{4}$

WRITTEN EXERCISE

Add:

1. $22\frac{2}{4}$	**2.** $16\frac{1}{2}$	**3.** $7\frac{3}{4}$	**4.** $8\frac{2}{3}$	**5.** $27\frac{3}{4}$
$5\frac{1}{4}$	$12\frac{1}{2}$	$9\frac{1}{4}$	$3\frac{2}{3}$	$4\frac{3}{4}$
$6\frac{3}{4}$	$4\frac{1}{2}$	$15\frac{3}{4}$	$19\frac{2}{3}$	$30\frac{3}{4}$

6. $\frac{3}{7}+\frac{6}{7}$

7. $1\frac{2}{3}+2\frac{1}{3}$

8. $9\frac{5}{6}+12\frac{5}{6}$

9. $14\frac{3}{4}+5\frac{3}{4}$

10. $11\frac{7}{8}+9\frac{7}{8}$

11. $4\frac{3}{4}+1\frac{3}{4}$

12. $5\frac{2}{9}+6\frac{8}{9}+2\frac{5}{9}$

13. $15\frac{2}{11}+1\frac{10}{11}$

14. $9\frac{5}{12}+20\frac{9}{12}+4\frac{3}{12}$

Subtract:

15. $2\frac{7}{9}-1\frac{5}{9}$

16. $8\frac{3}{11}-3\frac{3}{11}$

17. $7\frac{5}{12}-4\frac{4}{12}$

18. $4\frac{4}{7}-1\frac{1}{7}$

19. $5\frac{5}{3}-3$

20. $20-7\frac{1}{2}$

21. $36-21\frac{3}{4}$

22. $18\frac{1}{4}-8\frac{3}{4}$

23. A girl who had $\$7\frac{3}{4}$, put $\$5\frac{1}{4}$ in the school bank. How much did she keep on hand?

24. An automobile traveled $20\frac{2}{7}$ mi. on one day, $28\frac{3}{7}$ mi. the next, and $24\frac{1}{7}$ mi. another day. How far did it travel in the three days?

25. From $55\frac{8}{15}$ subtract $24\frac{7}{15}$.

Addition of Fractions with Unlike Denominators.

1. How many fourths in $\frac{1}{2}$? How many eighths in $\frac{1}{2}$? In $\frac{2}{2}$? In $\frac{2}{4}$? In $\frac{3}{4}$? In $\frac{5}{4}$?

2. Add $\frac{1}{2}$ and $\frac{1}{4}$. We change the denominator 2 to 4.

$$\frac{2}{4}+\frac{1}{4}=\frac{3}{4}.$$

3. Add $\frac{1}{2}$ and $\frac{3}{8}$. Here we must change 2 to what?

$$\frac{4}{8}+\frac{3}{8}=\frac{7}{8}.$$

4. Add $\frac{3}{4}$ and $\frac{5}{8}$. Here we must change 4 to what?

To add fractions with unlike denominators, first reduce them to fractions having a common denominator.

ORAL EXERCISE

Add:

1.	$\frac{1}{4}$	$\frac{1}{2}$	$\frac{1}{2}$	$\frac{3}{4}$	$\frac{1}{4}$	$\frac{1}{8}$	$\frac{5}{8}$	$\frac{1}{2}$
	$\frac{1}{8}$	$\frac{1}{4}$	$\frac{1}{8}$	$\frac{1}{8}$	$\frac{3}{8}$	$\frac{1}{2}$	$\frac{1}{4}$	$\frac{3}{8}$
2.	$\frac{1}{2}$	$\frac{3}{4}$	$\frac{1}{3}$	$\frac{1}{6}$	$\frac{2}{8}$	$\frac{4}{9}$	$\frac{1}{2}$	$\frac{5}{6}$
	$\frac{1}{12}$	$\frac{1}{12}$	$\frac{5}{12}$	$\frac{7}{12}$	$\frac{1}{12}$	$\frac{1}{12}$	$\frac{11}{12}$	$\frac{1}{12}$

WRITTEN EXERCISE

Add:

1.	2.	3.	4.	5.	6.
$5\frac{1}{4}$	$7\frac{1}{2}$	$8\frac{1}{2}$	$2\frac{3}{4}$	$3\frac{3}{4}$	$4\frac{1}{2}$
$2\frac{1}{4}$	$2\frac{1}{4}$	$11\frac{1}{2}$	$4\frac{1}{4}$	$3\frac{3}{4}$	$5\frac{1}{8}$

7. $\frac{1}{8}+\frac{1}{2}+\frac{3}{8}$
8. $\frac{1}{12}+\frac{7}{12}+\frac{11}{12}$
9. $\frac{5}{6}+\frac{5}{6}+\frac{1}{6}$
10. $\frac{8}{9}+\frac{7}{9}+\frac{5}{9}$
11. $\frac{1}{4}+\frac{1}{4}+\frac{3}{4}$
12. $\frac{5}{14}+\frac{9}{14}+\frac{7}{14}$
13. $\frac{7}{10}+\frac{3}{10}+\frac{9}{10}$
14. $\frac{5}{12}+\frac{7}{12}+\frac{1}{12}$
15. $\frac{2}{7}+\frac{6}{7}+\frac{4}{7}$
16. $\frac{2}{3}+\frac{2}{3}+\frac{1}{3}$
17. $\frac{7}{8}+\frac{7}{8}+\frac{5}{8}$
18. $\frac{4}{5}+\frac{3}{5}+\frac{3}{5}$

WRITTEN EXERCISE

1. Add $2\frac{1}{2}$, $4\frac{1}{4}$, and $3\frac{2}{6}$:

	12
$2\frac{1}{2}$	6
$4\frac{1}{4}$	3
$3\frac{2}{6}$	4
$10\frac{1}{12}$	$\frac{13}{12}=1\frac{1}{12}$

In addition of fractions arrange the work in this way. The 12 at the top shows the common denominator of 2, 4, and 6. Change the fractions to $\frac{6}{12}$, $\frac{3}{12}$ and $\frac{4}{12}$ and write 6, 3, and 4, under 12. The sum is $\frac{13}{12}$ or $1\frac{1}{12}$. Write $\frac{1}{12}$ and carry the 1 to the whole numbers.

Add:

2.	**3.**	**4.**	**5.**	**6.**
$2\frac{1}{8}$	$5\frac{1}{4}$	$8\frac{5}{8}$	$10\frac{1}{2}$	$30\frac{1}{4}$
$3\frac{5}{8}$	$7\frac{1}{2}$	$12\frac{3}{4}$	$18\frac{3}{8}$	$11\frac{1}{8}$
4	$8\frac{3}{8}$	$11\frac{1}{4}$	$2\frac{3}{4}$	$19\frac{1}{2}$

7.	**8.**	**9.**	**10.**	**11.**
$25\frac{5}{8}$	$28\frac{7}{8}$	$9\frac{7}{8}$	$19\frac{3}{4}$	$6\frac{3}{4}$
$31\frac{1}{2}$	$11\frac{1}{4}$	$26\frac{5}{8}$	$8\frac{1}{2}$	$17\frac{3}{8}$
$6\frac{3}{4}$	$12\frac{1}{2}$	$13\frac{1}{4}$	$14\frac{1}{8}$	$23\frac{1}{2}$

Find the sums:

12.	**13.**	**14.**	**15.**	**16.**
$3\frac{1}{2}$	$5\frac{1}{2}$	$1\frac{3}{4}$	$7\frac{1}{2}$	$8\frac{5}{6}$
$4\frac{1}{6}$	$1\frac{1}{4}$	$6\frac{1}{2}$	$2\frac{1}{3}$	$2\frac{1}{2}$
$1\frac{1}{6}$	$2\frac{1}{4}$	$3\frac{1}{5}$	$3\frac{3}{4}$	$1\frac{1}{4}$

17.	**18.**	**19.**	**20.**	**21.**
$3\frac{1}{2}$	$3\frac{3}{5}$	$4\frac{1}{5}$	$8\frac{3}{8}$	$9\frac{1}{2}$
$1\frac{5}{6}$	$4\frac{2}{10}$	$2\frac{3}{5}$	$9\frac{3}{4}$	$8\frac{3}{5}$
$1\frac{1}{4}$	$2\frac{1}{5}$	$1\frac{2}{10}$	$11\frac{1}{2}$	$3\frac{3}{8}$
2	1	$4\frac{3}{10}$	$9\frac{3}{4}$	4

22.	23.	24.	25.	26.
$9\frac{1}{6}$	$4\frac{2}{3}$	$6\frac{5}{8}$	$4\frac{1}{5}$	$4\frac{2}{3}$
$7\frac{5}{6}$	$3\frac{1}{6}$	$1\frac{1}{6}$	$3\frac{3}{8}$	$1\frac{5}{8}$
$2\frac{1}{2}$	$2\frac{7}{8}$	$1\frac{3}{4}$	$2\frac{3}{10}$	$3\frac{1}{6}$
$1\frac{1}{2}$	$4\frac{1}{2}$	5	9	$4\frac{3}{4}$

WRITTEN PROBLEMS

1. A clerk sold $2\frac{1}{4}$ yd. of muslin and $1\frac{1}{3}$ yd. of silk. How many yd. were sold in all?

2. He had $7\frac{2}{3}$ yd. of cotton goods in one piece and $4\frac{1}{3}$ yd. in another. How many yd. had he in both?

3. Your desk is $2\frac{3}{4}$ ft. long and the teacher's desk is $4\frac{1}{8}$ ft. long. Find the sum of the lengths.

4. One piece of molding is $1\frac{3}{8}$ in. wide and another $2\frac{1}{4}$ in. wide. What is the total width?

WRITTEN EXERCISE

Add:

1.	2.	3.	4.	5.
$26\frac{1}{2}$	$17\frac{3}{4}$	$93\frac{1}{6}$	$6\frac{3}{5}$	$19\frac{3}{8}$
$38\frac{2}{3}$	$2\frac{2}{3}$	$80\frac{4}{5}$	$19\frac{1}{4}$	$37\frac{3}{4}$
$8\frac{1}{6}$	$25\frac{1}{2}$	$22\frac{2}{3}$	$11\frac{3}{10}$	$2\frac{3}{12}$

6.	7.	8.	9.	10.
$8\frac{5}{6}$	$6\frac{7}{8}$	$27\frac{3}{8}$	$6\frac{3}{7}$	$7\frac{4}{9}$
$41\frac{3}{4}$	$8\frac{3}{4}$	$32\frac{2}{5}$	$7\frac{2}{3}$	$14\frac{2}{3}$
$26\frac{2}{3}$	$5\frac{1}{3}$	$9\frac{1}{10}$	$19\frac{1}{2}$	$30\frac{3}{4}$

11.	12.	13.	14.	15.
$2\frac{5}{18}$	$22\frac{3}{7}$	$42\frac{1}{2}$	$62\frac{5}{6}$	$22\frac{4}{9}$
$19\frac{1}{6}$	$33\frac{2}{3}$	$39\frac{4}{5}$	$30\frac{1}{2}$	$31\frac{1}{2}$
$17\frac{1}{2}$	$40\frac{2}{7}$	$51\frac{7}{10}$	$5\frac{1}{4}$	$16\frac{5}{6}$
$10\frac{1}{4}$	$15\frac{1}{2}$	$4\frac{3}{4}$	$7\frac{2}{3}$	$9\frac{1}{2}$

Add:

16.	**17.**	**18.**	**19.**	**20.**
$14\frac{9}{10}$	$37\frac{3}{16}$	$56\frac{1}{18}$	$75\frac{5}{12}$	$81\frac{5}{30}$
$22\frac{1}{2}$	$42\frac{1}{8}$	$32\frac{1}{2}$	$43\frac{3}{4}$	$11\frac{1}{2}$
$36\frac{3}{4}$	$11\frac{1}{2}$	$60\frac{5}{9}$	$30\frac{2}{9}$	$8\frac{5}{6}$
$40\frac{5}{8}$	$20\frac{3}{4}$	$12\frac{3}{4}$	$46\frac{1}{2}$	$7\frac{11}{15}$

ORAL EXERCISE

1. Reduce the following fraction to eighths:

$\frac{1}{2}$ $\quad\frac{1}{4}$ $\quad\frac{3}{4}$ $\quad\frac{4}{4}$ $\quad 1$ $\quad\frac{7}{2}$ $\quad\frac{4}{16}$ $\quad\frac{3}{24}$

2. Reduce to ninths:

$\frac{1}{3}$ $\quad 1$ $\quad\frac{2}{3}$ $\quad 3$ $\quad\frac{3}{3}$ $\quad\frac{4}{18}$ $\quad\frac{3}{27}$ $\quad 5$

3. Reduce to twelfths:

$\frac{3}{4}$ $\quad\frac{1}{3}$ $\quad\frac{1}{6}$ $\quad\frac{5}{6}$ $\quad\frac{4}{24}$ $\quad\frac{6}{6}$ $\quad\frac{4}{4}$ $\quad 2$

4. Reduce to sixteenths:

$\frac{3}{4}$ $\quad\frac{1}{4}$ $\quad\frac{1}{8}$ $\quad\frac{5}{8}$ $\quad\frac{4}{4}$ $\quad\frac{5}{8}$ $\quad\frac{7}{8}$ $\quad\frac{8}{8}$

5. Reduce to eighteenths:

$\frac{1}{2}$ $\quad\frac{1}{3}$ $\quad\frac{2}{3}$ $\quad 1$ $\quad\frac{3}{3}$ $\quad\frac{1}{6}$ $\quad\frac{5}{6}$ $\quad\frac{4}{9}$

6. Reduce to twentieths:

$\frac{1}{2}$ $\quad 1$ $\quad\frac{2}{2}$ $\quad 2$ $\quad\frac{1}{4}$ $\quad\frac{1}{5}$ $\quad\frac{3}{5}$ $\quad\frac{1}{10}$

7. Reduce to twenty-seconds:

$\frac{1}{2}$ $\quad\frac{3}{3}$ $\quad 1$ $\quad\frac{1}{11}$ $\quad\frac{3}{11}$ $\quad\frac{7}{11}$ $\quad\frac{10}{11}$ $\quad\frac{10}{10}$

8. Reduce to twenty-fifths:

$\frac{1}{5}$ $\quad 1$ $\quad\frac{3}{5}$ $\quad 3$ $\quad\frac{4}{5}$ $\quad\frac{10}{10}$ $\quad\frac{2}{5}$ $\quad\frac{5}{5}$

Subtraction of Fractions with Unlike Denominators. If a clerk cuts $2\frac{2}{3}$ yd. from a piece containing 12 yd., how many yd. will be left?

12	Since $\frac{3}{3}=1$, $12=11\frac{3}{3}$. $\frac{2}{3}+\frac{1}{3}=\frac{3}{3}$.
$2\frac{2}{3}$	Write $\frac{1}{3}$.
$9\frac{1}{3}$ yd.	$3+9=12$. Write 9.

ORAL EXERCISE

Subtract:

1.	$1\frac{1}{2}$	2	2	$10\frac{1}{2}$	1	1	$5\frac{3}{4}$	7
	$\frac{1}{2}$	$\frac{1}{2}$	$1\frac{1}{2}$	4	$\frac{1}{4}$	$\frac{2}{4}$	$\frac{2}{4}$	$\frac{3}{4}$
2.	$1\frac{1}{3}$	1	$5\frac{2}{3}$	$9\frac{2}{3}$	12	$3\frac{2}{3}$	$7\frac{2}{3}$	$5\frac{2}{3}$
	$\frac{1}{3}$	$\frac{1}{3}$	$\frac{1}{3}$	$7\frac{1}{3}$	$\frac{1}{3}$	$\frac{2}{3}$	$1\frac{1}{3}$	$5\frac{1}{3}$

WRITTEN EXERCISE

1. From 11 take $9\frac{3}{7}$:

11	
$9\frac{3}{7}$	Since $1=\frac{7}{7}$, $11=10\frac{7}{7}$. $\frac{3}{7}+\frac{4}{7}=\frac{7}{7}$. Write $\frac{4}{7}$.
	$10+1=11$. Write 1.
$1\frac{4}{7}$	

2. From $7\frac{5}{12}$ take $3\frac{3}{4}$:

	12	Here $\frac{9}{12}$ is greater than $\frac{5}{12}$.
$7\frac{5}{12}$	5	Since $1=\frac{12}{12}$, $7\frac{5}{12}=6\frac{5}{12}+\frac{12}{12}$ or $6\frac{17}{12}$.
$3\frac{3}{4}$	9	$\frac{9}{12}+\frac{8}{12}=\frac{17}{12}$. Write $\frac{8}{12}$ or $\frac{2}{3}$.
$3\frac{2}{3}$	$\frac{8}{12}=\frac{2}{3}$	$4+3=7$. Write 3.

Subtract:

3. 4	**4.** $8\frac{1}{2}$	**5.** $16\frac{3}{8}$	**6.** 21	**7.** $12\frac{1}{8}$
$1\frac{7}{8}$	$5\frac{3}{4}$	$11\frac{3}{4}$	$16\frac{3}{4}$	$7\frac{1}{4}$
8. $11\frac{3}{4}$	**9.** $25\frac{5}{6}$	**10.** 36	**11.** $28\frac{5}{6}$	**12.** $34\frac{2}{3}$
$7\frac{1}{5}$	$14\frac{3}{8}$	$12\frac{2}{7}$	$21\frac{1}{2}$	$27\frac{3}{10}$
13. $16\frac{2}{3}$	**14.** 43	**15.** $75\frac{5}{6}$	**16.** $56\frac{4}{5}$	**17.** $81\frac{3}{4}$
$8\frac{2}{5}$	$37\frac{7}{8}$	$68\frac{3}{8}$	$49\frac{3}{4}$	$72\frac{2}{3}$
18. $95\frac{7}{10}$	**19.** $80\frac{4}{9}$	**20.** $93\frac{7}{8}$	**21.** $87\frac{7}{8}$	**22.** $91\frac{5}{6}$
$18\frac{2}{5}$	$72\frac{1}{4}$	$86\frac{9}{10}$	$74\frac{3}{12}$	$35\frac{4}{5}$

ORAL REVIEW PROBLEMS

1. Ethel had a pound of tea and used $\frac{1}{4}$ of it. How much did she have left?

2. Last week mother had only $\frac{1}{4}$ pound of coffee. She bought $1\frac{1}{2}$ pounds. How much did she then have?

3. She baked 4 pies. We ate $1\frac{1}{2}$ pies. How many were left?

4. Suppose she had baked 3 pies. How many would have been left if we had eaten $1\frac{1}{3}$?

5. Father had a one-dollar bill and 3 quarters. He gave Van half a dollar. How much money did father have left?

6. Mother has a quarter of a pound of mustard in one can and $1\frac{1}{2}$ lb. in another. How much mustard has she?

7. We spent 1 hour in the room and $\frac{2}{3}$ of an hour in the yard. How much time did we spend in both?

8. Father had one hour and a quarter for lunch. Joe had only 30 minutes. How much more time had father than Joe?

9. How many inches are there in $\frac{1}{2}$ yd. of ribbon?

10. A pound of tea costs $.54 and a pound of coffee $\frac{5}{9}$ as much. How much will a pound of each cost?

11. What will $5\frac{1}{2}$ qt. of milk cost at 8¢ a qt.?

12. There are 48 members in the history club. Each member pays 25¢ a term dues. How much do the dues amount to?

13. What is the cost of $3\frac{1}{4}$ yd. muslin at 16¢ a yd.?

14. If I can read 15 pages of a book in 2 days, how long will it take me to finish 90 pages?

15. A boy buys magazines at $3\frac{1}{2}$¢ each and sells them at 5¢ each. How much will he make on a dozen copies?

WRITTEN EXERCISE

Add:

1.	2.	3.	4.	5.
$7\frac{3}{4}$	$14\frac{4}{7}$	$22\frac{1}{4}$	$9\frac{5}{6}$	$30\frac{1}{2}$
$8\frac{1}{2}$	$15\frac{1}{2}$	$13\frac{3}{8}$	$23\frac{3}{8}$	$46\frac{4}{5}$
$11\frac{5}{12}$	$9\frac{5}{14}$	$12\frac{7}{16}$	$34\frac{5}{12}$	$18\frac{7}{10}$

6.	7.	8.	9.	10.
$7\frac{3}{7}$	$16\frac{5}{8}$	$24\frac{3}{10}$	$4\frac{3}{4}$	$16\frac{5}{18}$
$24\frac{3}{14}$	$9\frac{2}{3}$	$8\frac{5}{6}$	$36\frac{11}{12}$	$22\frac{1}{9}$
$11\frac{1}{2}$	$3\frac{7}{12}$	$3\frac{4}{5}$	$41\frac{2}{3}$	$31\frac{2}{3}$
$5\frac{5}{14}$	$12\frac{1}{2}$	$17\frac{1}{2}$	$9\frac{1}{2}$	$8\frac{5}{6}$

Subtract:

11. $16\frac{1}{2}$	**12.** $34\frac{3}{4}$	**13.** $43\frac{1}{10}$	**14.** $59\frac{3}{8}$	**15.** $67\frac{3}{10}$
$3\frac{5}{6}$	$16\frac{7}{8}$	$27\frac{2}{5}$	$26\frac{5}{16}$	$49\frac{3}{5}$
16. $25\frac{3}{10}$	**17.** $37\frac{1}{6}$	**18.** $43\frac{3}{14}$	**19.** $58\frac{3}{7}$	**20.** $73\frac{1}{5}$
$11\frac{4}{5}$	$18\frac{3}{7}$	$12\frac{17}{28}$	$14\frac{3}{4}$	$11\frac{5}{6}$

21. $54\frac{3}{4}-49\frac{3}{8}$	**24.** $22\frac{1}{5}-8\frac{7}{8}$	**27.** $17\frac{5}{6}-12\frac{2}{7}$
22. $8\frac{11}{12}-7\frac{3}{4}$	**25.** $8\frac{2}{3}-6\frac{1}{9}$	**28.** $14\frac{4}{7}-6\frac{2}{5}$
23. $9\frac{5}{8}-7\frac{1}{2}$	**26.** $8\frac{7}{12}-1\frac{1}{5}$	**29.** $6\frac{1}{9}-2\frac{1}{5}$

The Least Common Multiple: Advanced Work. The product of two or more integers is called a **multiple** of each of them. 12 is a multiple of 2, 3, 4, and 6 because $2\times2\times3=3\times4=2\times6=12$.

A multiple of two or more numbers is called a **common multiple** of the numbers. 20 is a common multiple of 5 and 4, and so are 40, 80, 100, etc.

The smallest common multiple of two or more numbers is called their **least common multiple** (L. C. M.). It is the least number that is exactly divisible by each of them. 80 is a common multiple of 5 and 4, but their least common multiple is 20.

A **prime factor** is a factor which is exactly divisible by no number except itself and one.

We find the L. C. M. of numbers by factoring them and finding the product of all the prime factors, using each factor the greatest number of times it occurs in any number.

Find the L. C. M. of 30 and 45:

The prime factors of 30 are 2, 3, and 5. Notice that 2 occurs only once as a factor of 30; likewise 3; likewise 5. The prime factors of 45 are 5, 3, and 3. Notice that 3 occurs twice as a factor of 45, while 5 occurs only once. Therefore in finding the L. C. M. of 30 and 45, we use 3 twice and the other two factors each once.

Hence, L. C. M. of 30 and 45 $= 5 \times 2 \times 3 \times 3 = 90$.

WRITTEN EXERCISE

Find the L. C. M. of:

1.	2.	3.	4.
16 and 24	15 and 35	27 and 39	21 and 36
33 and 55	28 and 35	18 and 48	6, 15, 39

A Fraction Multiplied by an Integer. How much is 2 times $\frac{1}{3}$? 4 times $\frac{1}{4}$? 5 times $\frac{3}{25}$?

To multiply a fraction by an integer, multiply the numerator by the integer.

ORAL EXERCISE

Multiply:

1.	2.	3.	4.	5.	6.
$\frac{2}{3}\times 4$	$\frac{4}{5}\times 2$	$\frac{3}{7}\times 5$	$\frac{7}{8}\times 4$	$\frac{6}{7}\times 3$	$\frac{2}{3}\times 18$
$\frac{7}{9}\times 81$	$\frac{1}{3}\times 96$	$\frac{5}{8}\times 48$	$\frac{5}{9}\times 27$	$\frac{4}{10}\times 70$	$\frac{6}{7}\times 84$
$\frac{3}{5}\times 35$	$\frac{5}{12}\times 36$	$\frac{4}{7}\times 49$	$\frac{5}{6}\times 24$	$\frac{7}{8}\times 32$	$\frac{3}{11}\times 33$
$\frac{4}{5}\times 6$	$\frac{2}{9}\times 7$	$\frac{3}{8}\times 4$	$\frac{2}{3}\times 9$	$\frac{2}{7}\times 5$	$\frac{3}{11}\times 4$

An Integer Multiplied by a Fraction. To take a fractional part of a number is called *multiplication by a fraction.* To take $\frac{1}{4}$ of 12 is called the multiplication of 12 by $\frac{1}{4}$.

ORAL EXERCISE

1.	2.	3.	4.	5.
$\frac{2}{3}$ of 12	$\frac{7}{12}$ of 48	$\frac{4}{5}$ of 20	$\frac{2}{5}$ of 65	$\frac{5}{6}$ of 120
$\frac{5}{8}$ of 40	$\frac{7}{8}$ of 64	$\frac{7}{12}$ of 72	$\frac{7}{8}$ of 96	$\frac{2}{3}$ of 350

A Fraction Multiplied by a Fraction.

1. What is $\frac{1}{2}$ of $\frac{1}{2}$ of this page? What is $\frac{1}{4}$ of $\frac{1}{2}$? What is $\frac{1}{5}$ of $\frac{1}{2}$?

2. Make drawing to prove your answers.

3. Find $\frac{3}{4}$ of $\frac{16}{27}$: $\frac{\cancel{3}}{\cancel{4}}$ of $\frac{\overset{4}{\cancel{16}}}{\underset{9}{\cancel{27}}} = \frac{4}{9}$.

To multiply a fraction by a fraction, multiply the numerators together for a new numerator, and the denominators together for a new denominator. Use cancellation wherever possible.

WRITTEN EXERCISE

Multiply:

1. $\frac{1}{2}$ of $\frac{1}{3}$	6. $\frac{2}{3}$ of $\frac{4}{5}$	11. $\frac{5}{8}$ of $\frac{3}{10}$
2. $\frac{1}{3}$ of $\frac{3}{8}$	7. $\frac{6}{7}$ of $\frac{3}{4}$	12. $\frac{6}{7}$ of $\frac{5}{12}$
3. $\frac{1}{2}$ of $\frac{2}{9}$	8. $\frac{2}{3}$ of $\frac{7}{8}$	13. $\frac{3}{10}$ of $\frac{7}{90}$
4. $\frac{3}{5}$ of $\frac{1}{2}$	9. $\frac{2}{5}$ of $\frac{15}{18}$	14. $\frac{3}{5}$ of $\frac{15}{60}$
5. $\frac{1}{2}$ of $\frac{4}{5}$	10. $\frac{3}{4}$ of $\frac{7}{24}$	15. $\frac{5}{6}$ of $\frac{37}{40}$

A Fraction Divided by an Integer.

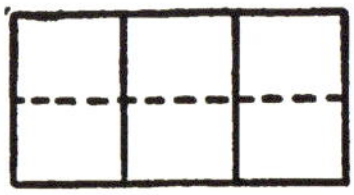

1. Divide $\frac{1}{3}$ of this rectangle by 2. What does each part equal? What does $\frac{1}{2}$ of $\frac{1}{3}$ equal? $\frac{1}{3}\div 2=\frac{1}{3}\times\frac{1}{2}=\frac{1}{6}$.

To divide a fraction by an integer, multiply the denominator by the integer.

2. Divide $\frac{2}{3}$ of the rectangle by 2. What does each part equal? What does $\frac{2}{3}\div 2$ equal? What is $\frac{1}{2}$ of $\frac{2}{3}$?

$$\frac{2}{3}\div 2=\frac{2\div 2}{3}=\frac{1}{3}.$$

To divide a fraction by an integer, divide the numerator by the integer.

WRITTEN EXERCISE

Divide:

1. $\frac{7}{8}\div 7$	5. $\frac{1}{2}\div 9$	9. $\frac{6}{11}\div 2$	13. $\frac{2}{7}\div 4$
2. $\frac{2}{5}\div 3$	6. $\frac{12}{17}\div 3$	10. $\frac{3}{4}\div 2$	14. $\frac{8}{9}\div 4$
3. $\frac{1}{3}\div 4$	7. $\frac{4}{13}\div 2$	11. $\frac{5}{8}\div 4$	15. $\frac{6}{7}\div 2$
4. $\frac{1}{6}\div 5$	8. $\frac{8}{15}\div 4$	12. $\frac{7}{8}\div 5$	16. $\frac{3}{5}\div 6$

An Integer Divided by a Fraction.

1. How many quarters of an inch in 1 inch? In 3 inches? In 6 inches?

2. $1\div\frac{1}{4}=?$ $1\div\frac{1}{4}=1\times 4=4.$ $3\div\frac{1}{4}=?$ $6\div\frac{1}{4}=?$

3. $6\div\frac{3}{4}=?$ If $6\div\frac{1}{4}=6\times 4$, $6\div\frac{3}{4}=\frac{\overset{2}{\cancel{6}}\times 4}{\cancel{3}}=8.$

To divide an integer by a fraction, invert the fraction used as a divisor and multiply.

WRITTEN EXERCISE

Divide:

1. $8 \div \frac{1}{4}$	5. $20 \div \frac{3}{4}$	9. $40 \div \frac{4}{7}$	13. $13 \div \frac{1}{6}$
2. $12 \div \frac{1}{3}$	6. $18 \div \frac{6}{7}$	10. $25 \div \frac{5}{8}$	14. $30 \div \frac{1}{7}$
3. $6 \div \frac{2}{3}$	7. $31 \div \frac{1}{2}$	11. $33 \div \frac{3}{4}$	15. $36 \div \frac{3}{8}$
4. $8 \div \frac{2}{9}$	8. $38 \div \frac{2}{5}$	12. $50 \div \frac{1}{2}$	16. $28 \div \frac{2}{3}$

17. Into how many pieces each $\frac{2}{3}$ of a yd. long can 24 yd. of ribbon be cut?

18. A strip of carpet 50 yards long was cut into mats each $\frac{5}{8}$ of a yard in length. How many mats were there?

A Fraction Divided by a Fraction. What does $\frac{2}{3} \div \frac{3}{4}$ equal? We have seen that $1 \div \frac{3}{4} = 1 \times \frac{4}{3} = \frac{4}{3}$.

$$\frac{2}{3} \div \frac{3}{4} = \frac{2}{3} \times \frac{4}{3} = \frac{8}{9}.$$

To divide a fraction by a fraction, invert the divisor and multiply.

WRITTEN EXERCISE

Divide:

1. $\frac{2}{3} \div \frac{2}{5}$	5. $\frac{9}{10} \div \frac{90}{100}$	9. $\frac{3}{16} \div \frac{5}{8}$	13. $\frac{7}{20} \div \frac{5}{7}$
2. $\frac{7}{8} \div \frac{3}{4}$	6. $\frac{6}{7} \div \frac{42}{54}$	10. $\frac{7}{8} \div \frac{3}{4}$	14. $\frac{5}{12} \div \frac{20}{24}$
3. $\frac{4}{5} \div \frac{3}{4}$	7. $\frac{4}{5} \div \frac{2}{3}$	11. $\frac{9}{10} \div \frac{5}{7}$	15. $\frac{8}{9} \div \frac{32}{35}$
4. $\frac{8}{9} \div \frac{40}{45}$	8. $\frac{3}{8} \div \frac{9}{11}$	12. $\frac{7}{15} \div \frac{3}{5}$	16. $\frac{5}{8} \div \frac{25}{27}$

17. How many sashes each $2\frac{5}{8}$ yd. long can be made from $12\frac{1}{4}$ yd. of ribbon?

18. If it takes $1\frac{1}{3}$ yd. of flannel to make a school pennant, how many pennants can be cut from $81\frac{1}{3}$ yd.?

ORAL DRILL EXERCISE

Find answers:

	A	*B*	*C*	*D*	*E*
1.	$\frac{1}{2}+\frac{3}{4}$	$\frac{1}{4}+\frac{3}{8}$	$\frac{3}{4}+\frac{3}{8}$	$\frac{1}{4}+\frac{3}{4}$	$\frac{1}{4}+\frac{5}{8}$
2.	$2\frac{1}{8}+2\frac{1}{2}$	$3\frac{1}{4}+1\frac{1}{2}$	$4\frac{3}{8}+2\frac{5}{8}$	$1\frac{7}{8}+1\frac{1}{2}$	$2\frac{3}{8}+4\frac{1}{4}$
3.	$1-\frac{1}{8}$	$1-\frac{3}{8}$	$2-\frac{1}{4}$	$3-\frac{3}{4}$	$2-\frac{2}{3}$

Reduce to lowest terms:

4.	$\frac{2}{4}$	$\frac{3}{6}$	$\frac{6}{8}$	$\frac{4}{10}$	$\frac{8}{12}$
5.	$\frac{3}{3}$	$\frac{2}{6}$	$\frac{6}{9}$	$\frac{4}{8}$	$\frac{8}{10}$
6.	$\frac{2}{12}$	$\frac{4}{8}$	$\frac{4}{12}$	$\frac{2}{10}$	$\frac{2}{8}$
7.	$\frac{2}{14}$	$\frac{10}{12}$	$\frac{8}{4}$	$\frac{20}{12}$	$\frac{8}{6}$
8.	$\frac{10}{8}$	$\frac{12}{10}$	$\frac{10}{10}$	$\frac{18}{8}$	$\frac{9}{8}$

State results:

9.	$\frac{2}{3}\times 45$	$\frac{4}{5}\times 65$	$\frac{3}{4}\times 80$	$\frac{7}{8}\times 96$	$\frac{5}{6}\times 72$
10.	$\frac{1}{3}+\frac{1}{3}+\frac{1}{3}$	$\frac{1}{2}+\frac{3}{4}+\frac{1}{2}$	$\frac{1}{4}+\frac{3}{4}+\frac{1}{4}$	$\frac{2}{3}+\frac{2}{3}+\frac{1}{3}$	$\frac{2}{3}+\frac{2}{3}+\frac{2}{3}$
11.	$5\frac{2}{5}=\frac{\ }{5}$	$4\frac{2}{3}=\frac{\ }{3}$	$7\frac{1}{8}=\frac{\ }{8}$	$5\frac{2}{7}=\frac{\ }{7}$	$8\frac{3}{8}=\frac{\ }{8}$

Add:

12.	$5.60 .44	$11.40 .56	$3.80 .33	$9.20 .78	$6.80 .25
13.	$7.20 .89	$12.90 .72	$4.70 .49	$12.40 .39	$9.10 .95

VIII. MEASURES

Area of Rectangles. A **rectangle** is a figure having four sides and four square corners. The blackboard, your ruler, the window, the door are rectangles.

When the sides of the rectangle are equal, we have a **square**. The area of a square that is 1 inch on a side is 1 square inch.

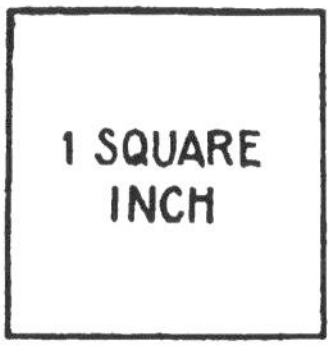

1. Draw a rectangle 2 inches long and 1 inch wide. What is its area?

2. Draw a square 3 inches long. How wide is it? What is its area?

3. Draw a square 5 inches long. How wide is it? Find its area.

4. Draw a rectangle 4 inches long and 3 inches wide. Find its area by dividing it into square inches.

5. How many square inches are there in a book cover 10 inches long and 7 inches wide?

6. Draw a rectangle 8 inches long, making its width one half its length. How many square inches does it contain?

We find the area of a rectangle by multiplying the number of inches in the length by the number of inches in the width.

Table of Square Measure.

144 square inches (sq. in.)	= 1 square foot (sq. ft.)
9 square feet	= 1 square yard (sq. yd.)

ORAL EXERCISE

1. A paper is 8 inches long and 5 inches wide. Find the area.

2. A book is 5 inches long and its area is 40 square inches. How wide is it?

3. A board is 2 yards long and 3 yards wide. Find its area. Another board is 2 yards long and 2 feet wide. Find its area.

4. The area of a rectangle is 28 square inches. The width is 4 inches. Find the length. Another rectangle is 3 yards long, with an area of 48 square yards. Find its length.

5. A board is 12 ft. long and 10 in. wide. Find its area in sq. in.

6. A playground is 16 yd. long and 6 yd. wide. Find its area in sq. yd.

WRITTEN EXERCISE

Find the area of each of these rectangles:

1. Length 16 inches, width 9 inches.
2. Length 20 inches, width 11 inches.
3. Length 12 feet, width 6 ft.
4. Length 8 yards, width 12 yd.

Find the width of these rectangles:

5. Length 11 inches, area 66 sq. in.
6. Length 8 feet, area 48 sq. ft.
7. Length 3 yards, area 33 sq. yd.
8. Length 9 ft., area 81 sq. ft.

Find the length of these rectangles:

9. Width 7 in., area 84 sq. in.
10. Area 55 sq. ft., width 5 feet.
11. Width 4 yd., area 24 sq. yd.
12. Area 16 sq. yd., width 2 yd.

Finding the Perimeter. If our room is 30 feet long and 22 feet wide, how many feet of picture molding will be needed for it?

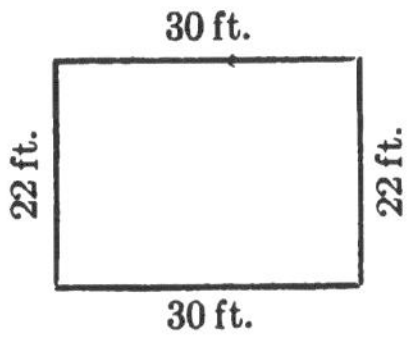

The distance around a figure is called the **perimeter**.

WRITTEN EXERCISE

Find the perimeter of each of these rectangles:

1. 12 in. by 8 in.
2. 10 yd. by 9 yd.
3. 15 in. by 11 in.
4. 12 feet by 6 feet.
5. 11 yd. by 4 yd.
6. $2\frac{1}{2}$ yd. by 1 yd.

Find the length of each of the following:

7. The perimeter is 30 in.; the width is 5 in.
8. The perimeter is 24 in.; the width is 6 in.
9. The perimeter is 28 ft.; the width is 7 ft.
10. The perimeter is 50 yd.; the width is 10 yd.

WRITTEN REVIEW PROBLEMS

1. A wall is 49 inches wide and 54 inches long. How many square inches does it contain?

2. Measure the length and width of your classroom and find its area. Find the area of two rooms in your house.

3. A square is 5 yards on a side. Find its perimeter and area.

4. The area of a yard is 612 sq. yd. It is 17 yd. in length. Find the width of the yard.

5. Find the perimeter of this yard.

6. How many feet of wire fencing would be needed to enclose it?

7. Find the perimeter of a room $18\frac{1}{4}$ ft. long and $11\frac{1}{4}$ ft. wide.

8. The perimeter of a lot is 394 ft. The width is 25 ft. Find the length.

9. Find the area of this lot in sq. ft.

10. A farmer's field runs back 145 feet from the road. 940 feet of fence surround it. How far along the road does it run? Make a diagram of this field.

11. A room is 32 ft. by 21 ft. How many square feet of carpet will be needed to cover it?

12. Picture molding is 8¢ a foot. How much will it cost to run a molding around this room?

13. How many sq. ft. of sod will it take to cover a lot 43 ft. wide by 155 ft. long?

Cubic Measure: Advanced Work.

What is the length of each edge of this one-inch cube? If we made a cube 2 inches long, 2 inches wide, and 2 inches high, how many one-inch cubes would there be in it? If we made one 3 inches long, 3 inches wide, and 3 inches high, how many 1-inch cubes would there be in it?

To find the volume of a solid, we multiply the number of inches, feet, or yards in its length, by the number of the same denomination in its width, and this product by the number of the same denomination in its thickness.

Table of Cubic Measure.

1728 cubic inches (cu. in.)	=1 cubic foot (cu. ft.)
27 cubic feet	=1 cubic yard (cu. yd.)

1. A rectangle is 3 in. long and 5 in. wide. How many one-inch cubes can I place upon it?

If I pile these 3 layers deep, how many cubic inches will the pile contain?

2. How many cubic feet in a piece of stone 6 ft. long, 10 ft. wide, 3 ft. deep?

3. Find the volume of a box 8 ft. long, 5 ft. wide, and 4 ft. deep.

4. Find the volume of a tank, 9 yd. long, 3 yd. wide, and 2 yd. deep.

5. How many cu. yd. in a room 10 yd. long, 5 yd. high, and 4 yd. wide?

Compound Numbers.

1. One line is 9 in. long, and another is 11 in. How long are the two lines?

2. A boy spent 25 min. in one room and 1 hr. 40 min. in another. How much time did he spend in both rooms?

A **compound number** is one that is expressed in two or more denominations. 1 hr. 40 min. is a compound number.

Addition of Compound Numbers.

Add 8 bu. 3 pk. and 22 bu. 2 pk.:

bu.	pk.
8	3
22	2
31	1

Since we have 5 pk., we write 1 pk. and change 4 pk. to 1 bu. We add the 1 bushel to the 30, making 31.

WRITTEN EXERCISE

Add:

1.

hr.	min.
2	40
16	35

2.

gal.	qt.
12	3
9	2

3.

gal.	qt.	pt.
3	3	1
4	3	1

Add the following:

	4.			5.			6.	
hr.	min.	sec.	pk.	qt.	pt.	cwt.	lb.	oz.
16	45	32	2	7	1	3	54	8
4	18	40	0	3	0	2	97	14

	7.			8.			9.	
yd.	ft.	in.	da.	hr	min.	T.	lb.	oz.
4	2	8	2	20	40	1	1084	11
7	2	9	0	18	45	1	997	14

	10.			11.			12.	
mi.	yd.	ft.	bu.	pk.	qt.	gal.	qt.	pt.
1	1200	2	1	3	7	7	3	1
1	860	2	3	3	5	3	2	1

Subtraction of Compound Numbers.

1. A grocer had 4 lb. of tea. How much had he left after selling 1 lb. 12 oz.?

2. One line is 3 ft. 10 in. long. Another is 1 ft. 8 in. in length. Find the difference in lengths.

3. From 22 ft. 4 in. take 18 ft. 7 in.:

ft.	in.	
22	4	Here we reduce 22 ft. 4 in. to 21 ft. 16 in.
18	7	7+9=16. Write 9.
3	9	19+3=22. Write 3.

WRITTEN EXERCISE

Subtract:

	1.			2.			3.	
hr.	min.	sec.	T.	lb.	oz.	hr.	min.	sec.
7	14	35	17	284	3	16	17	40
5	25	30	3	200	11	10	30	50

	4.			5.			6.	
pk.	qt.	pt.	gal.	qt.	pt.	bu.	pk.	qt.
3	3	0	8	3	1	18	3	2
1	5	1	6	1	1	7	0	7

	7.			8.			9.	
yd.	ft.	in.	yr.	mo.		da.	hr.	min.
17	2	3	7	2		3	7	15
6	1	5	2	10		1	14	45

	10.			11.			12.	
bu.	pk.	qt.	gal.	qt.	pt.	pk.	qt.	pt.
7	3	1	7	3	0	8	3	0
3	1	5	5	3	1	2	6	1

GENERAL ORAL PROBLEMS

1. From a piece of goods containing 15 yd., $\frac{2}{5}$ were cut off. The remainder was sold at 10¢ a yd. How much was it sold for?

2. $\frac{1}{8}$ of a yd. of silk costs 25¢. Find the cost of 1 yard.

3. One piece of cloth contained 38 yards; another contained $57\frac{1}{4}$ yards. How many yards were there in both pieces?

4. The grocer sold a doz. boxes of matches with 60 matches in each box. How many matches did he sell?

5. A farmer bought a cow for $42 and sold it at a gain of $19. How much did he receive for it?

6. I sold an article for $12.50. It cost $15. Find the loss.

7. I gained $.70 on a fountain pen. It cost $2.25. Find the selling price.

8. I lost $1.20 on a pair of skates. They cost $5.00. Find the selling price.

9. Sadie bought 3 pieces of ribbon. Each contained $6\frac{1}{2}$ yards. How many yards had she?

10. Helen bought $\frac{1}{4}$ lb. butter at 36¢ and $\frac{1}{2}$ lb. steak at 42¢. How much money did she spend?

11. If 9 hats cost $5, how many can I buy for $20?

12. Collars are 2 for a quarter. What will a dozen cost?

13. If 1000 bricks cost $6, find the cost of 4500.

14. A dealer had 96 pairs of gloves and sold $\frac{1}{8}$ of them at $1.25 a pair. How much did he receive for them?

15. A yard is 20 ft. long and 17 ft. wide. How many feet of fence will be needed to enclose it?

16. I bought $2\frac{1}{2}$ lb. of bacon at 24¢. How much change did I receive from a dollar bill?

ORAL DRILL EXERCISE

	A	*B*	*C*	*D*
1.	$\frac{15\times3}{10}$	$\frac{3}{16}$ of 32	$\frac{6}{7}$ of 49	$\frac{7}{8}$ of 104
2.	$3\frac{1}{2}=\frac{\ }{2}$	$\frac{36\times4}{6}$	$9\frac{2}{3}=\frac{\ }{3}$	$\frac{8\times6}{4}$
3.	$\frac{8}{9}$ of 81	4)70	$\frac{4}{9}$ of 54	11)100
4.	$\frac{64\times4}{8}$	$4\frac{4}{5}=\frac{\ }{5}$	$\frac{28\times5}{7}$	$12\frac{1}{2}=\frac{\ }{2}$
5.	$9\frac{1}{7}=\frac{\ }{7}$	$\frac{3}{20}$ of 100	$8\frac{2}{3}=\frac{\ }{3}$	$\frac{5}{8}$ of 64
6.	$\frac{3}{10}$ of 90	$\frac{4}{9}$ of 81	$\frac{4}{5}$ of 80	13×8
7.	$70\div5$	18×5	$85\div5$	$\frac{3}{8}$ of 40
8.	$22+88-12$	$75\div5$	$\frac{6}{7}$ of 49	$47+95-17$
9.	$\frac{2}{3}$ of 36	$\frac{5}{12}$ of 72	$67+34-13$	$\frac{3}{4}$ of 48
10.	14×7	$49+21-18$	16×6	$65\div5$

	Find areas:		*State results:*		*Find areas:*	
	LENGTH	WIDTH			LENGTH	WIDTH
11.	10 in.	12 in.	5)63	12×200	12 in.	8 in.
12.	18 ft.	10 ft.	9)80	30×400	8 yd.	24 yd.
13.	22 yd.	8 yd.	6)59	22×50	33 ft.	5 ft.
14.	60 yd.	4 yd.	12)90	77×20	26 ft.	10 ft.
15.	7 in.	12 in.	7)45	90×80	20 yd.	14 yd.
16.	7 ft.	13 ft.	8)60	14×80	13 in.	30 in.

Subtract:

17.	$18\frac{1}{2}$ $-2\frac{1}{4}$	$29\frac{3}{4}$ $-5\frac{1}{2}$	$33\frac{1}{2}$ $-4\frac{1}{4}$	$42\frac{1}{4}$ $-7\frac{1}{4}$
18.	6 $-\frac{1}{4}$	7 $-\frac{1}{2}$	9 $-\frac{3}{4}$	11 $-\frac{1}{3}$

GENERAL WRITTEN PROBLEMS

1. A storekeeper bought 38 yd. of cloth for $74.50 and sold it at $1.75 a yd. Find how much he gained or lost.

2. Find the cost of covering with tin a roof 48 feet by 32 feet at $.25 a sq. ft.

3. 20 barrels of cranberries were sold for $135. What would 31 barrels cost at this rate?

4. An agent is given $42,750 to purchase farm horses. How many can he buy at $95 each?

5. A man had 175 bushels of grain; $58\frac{1}{2}$ bu. were barley, $44\frac{3}{4}$ bu. were oats, $35\frac{1}{8}$ rye, and the remainder wheat. How many bushels of wheat did he have?

6. Mother paid $2.64 for the following: 3 lb. of tea at $.60, and 14 lb. of sugar. How much per pound did she pay for sugar?

7. A party of Boy Scouts marched $9\frac{2}{3}$ miles on Monday, $11\frac{3}{5}$ miles on Tuesday, and $15\frac{4}{6}$ miles on Wednesday. How much farther have they to go to finish a trip of 40 miles?

8. An automobile was run $6\frac{2}{3}$ hours one day, $4\frac{1}{[illegible]}$ hours a second day, and $8\frac{1}{2}$ hours another day. How much longer will it have to go in order to have traveled 20 hours in all?

9. An aeroplane was making a trip of $115\frac{4}{5}$ miles. After traveling $89\frac{1}{3}$ miles it was forced to come down. How far had it still to fly in order to complete the trip?

10. A kitchen is 4 yd. square. Find the cost of covering the floor with oilcloth at $1.05 a sq. yd.

11. A billboard is 3 yards high and 48 yards long. How many square yards does it contain?

12. How many yards of metal border will be needed to go around it? This metal border is sold by the foot. How many feet of it will be needed?

13. A storekeeper's gas bill for one year amounted to $220.68. Find the average monthly bill.

14. A factory can turn out 270 suits in 15 days. How many days will be required to turn out 828 suits?

15. Find the cost of $4\frac{1}{2}$ doz. cans of pineapples at $2.10 a dozen cans.

16. What must I pay for $1\frac{1}{2}$ doz. cans of spinach at 11¢ a can?

17. A man receives $2912 each year. How much does he receive every week?

18. A dealer bought 86 sets of furniture and paid a bill of $2064. Find the price of each set.

19. How many cars will be required to carry 3591 tons of stone, if each car holds 59 tons?

20. One bit of rope was $8\frac{1}{2}$ ft. long. Another was $5\frac{2}{3}$ ft. long. How long were both?

21. A grocer charges 6¢ a pound for sugar and 45¢ a dozen for eggs. At a special sale a customer buys $4\frac{1}{2}$ pounds of sugar for 19¢ and $2\frac{1}{2}$ dozen eggs for 95¢. How much does she save on her bill?

IX. BILLS AND RECEIPTS

What a Bill Shows. In the following bill read the name and address of the purchaser; the name and address of the person who sold the goods; the date of the sale; the date the bill was made out; the items, or the articles sold; the amount of each sale; the footing, or total, of the bill; and the receipt showing when and by whom the money was received.

NEW YORK, May 1, 1917.

Mr. Frank Hardy,
141 East 15th St.,

Bought of GEORGE H. LOUIS & CO., Grocers,
225 FOURTH AVENUE.

Apr. 24	8 lb. coffee	@ .35	2	80
	3 doz. eggs	.38	1	14
	16 lb. sugar	.08	1	28
		RECEIVED PAYMENT MAY 5, 1917 GEORGE H. LOUIS & CO. *Per C. W. A.*	5	22

1. In the above bill, check the items to see if they are charged correctly.

2. Has this bill been paid? Who is C. W. A.?

WRITTEN EXERCISE

Make out bills for goods purchased, as follows. Use the names of dealers in your neighborhood, and the names of your classmates as purchasers. Receipt each bill, using your initials as clerk.

1. 8 yd. silk @ \$2.25; 12 yd. flannel @ \$.25; 24 yd. linoleum @ \$1.55.

2. $\frac{1}{2}$ bu. potatoes @ \$.90; 15 lb. sugar @ \$.06; 8 lb. tea @ \$.38.

3. 4 gal. paint @ \$1.40; 4 brushes @ \$.65; 1 gal. turpentine @ \$.20.

4. 6 chairs @ \$2.35; 1 mirror @ \$9.50; 1 couch @ \$18.90; 1 armchair @ \$7.65.

5. 6 doz. oranges @ \$.48; 1 doz. boxes matches @ \$.03; 10 lb. butter @ \$.36; 9 boxes crackers @ \$.10.

6. $3\frac{1}{2}$ lb. round steak @ \$.24; 14 lb. ham @ \$.21; $7\frac{1}{2}$ lb. bacon @ \$.20.

7. 2 hammers @ \$.85; 12 boxes tacks @ \$.08; 2 doz. pairs hinges @ \$3.80; 1 doz. locks @ \$6.80.

8. 28 lb. tea @ \$.48; $9\frac{1}{2}$ lb. grapes @ \$.09; $3\frac{1}{4}$ lb. butter @ \$.44.

9. 12 grape fruit @ \$.08; $1\frac{1}{4}$ doz. oranges @ \$.52; 3 doz. bananas @ \$.18.

10. $6\frac{3}{4}$ lb. fish @ \$.28; $2\frac{1}{3}$ lb. steak @ \$.30; $1\frac{1}{2}$ lb. lard @ \$.18.

11. 2 doz. pencils @ \$.55; $\frac{1}{2}$ doz. erasers @ \$.80; $8\frac{1}{4}$ doz. pads @ \$.60.

Receipts. In the following receipt read the name of the person who pays the money, the person who receives it, the date of payment, the amount paid:

$724 $\frac{50}{100}$	***New York, May 14, 1917.***
Received of......*John H. Dobson*................	
Seven hundred twenty-four ~~~~~~~~~~ $\frac{50}{100}$ **Dollars**	
for services rendered.	
	Charles L. Long.

WRITTEN EXERCISE

Make out receipts for the following:

1. On August 4, 1916, George H. Hickey paid William Kneissel $208.40 for services rendered.

2. On January 18, 1917, Martha Robinson paid Dr. Charles Devlin $45 for professional services.

3. On May 1, 1917, Esther Green paid Barrett, Linden & Co. $140.62 for services rendered.

4. On February 6, 1914, John I. Trimble paid R. H. Carey & Co. $87.80 for services rendered.

5. On Dec. 4, 1917, Ruth Benson paid Dr. Harriet Kirby $60 for professional services.

6. Make out separate receipts for the bills on page 163.

7. A boy from A. G. Spalding & Bros. delivers to you today 4 basket balls for the school. Write out a receipt for him.

WRITTEN REVIEW PROBLEMS

1. Mother is making hats. She used $2\frac{3}{4}$ yd. ribbon on one, $1\frac{5}{8}$ yd. ribbon on another, and $3\frac{2}{5}$ yd. on the last. How many yd. of ribbon did she use?

2. From a piece of cotton goods containing 83 yd., a salesman sold $15\frac{2}{3}$ yd., $26\frac{1}{2}$ yd., and $12\frac{1}{5}$ yd. How many yd. were left?

3. The first part of a book contains 180 pages. The second part has $\frac{3}{5}$ as many pages as the first. How many pages in the book?

4. If $\frac{1}{2}$ yd. of carpet cost \$2.50, what will 40 yd. cost?

5. A plumber needs $3\frac{3}{4}$ ft. of pipe for one job, $12\frac{2}{3}$ ft. for another, and $26\frac{7}{12}$ ft. for a third job. How many feet of pipe must he buy?

6. Find the area of a field 124 feet long and 79 feet wide.

7. A clerk sold $5\frac{3}{5}$ yd. of muslin, $12\frac{1}{3}$ yd. of silk, and as many yd. of cotton goods as the sum of the other two amounts. How many yd. were sold in all?

8. A moving picture screen is 16 feet high and $10\frac{1}{2}$ feet wide. Find its area.

9. A metal border running all around it is worth \$.44 a foot. Find the cost of the border.

10. Another screen contains 195 sq. ft. Its length is 15 ft. Find the width.

11. Find the difference in area of the two screens.

12. A mirror contains 1440 square inches. One side is 40 inches. Find the other.

13. If oranges are 3¢ each, how many dozen can be bought for $15.48?

14. A man earns $1200 a year. If his expenses are $576 yearly, in how many years can he save $7488?

15. A man bought an automobile for $750. If he agreed to pay $230 down and $40 a month afterward, in how many months would he have paid all he owed?

16. The expenses of a family of four persons were $2172 for a year. Find the average monthly expenses of each person.

17. A wood lot containing 12 acres was sold for $2544. A wheat farm of 62 acres was sold for $6820. Find the difference in the price per acre.

18. A chauffeur bought 22 gal. of gasoline for $5.72. If he buys 6 gal. more, how much will it all cost?

19. A shipload of 950 cattle was sold for $39,900. How much was obtained for each head of cattle?

20. If a clerk earns $137.50 a month and spends $114.40, how much can he save in 2 years?

21. How many inches of ribbon in three pieces containing $\frac{1}{2}$ yd., $3\frac{1}{2}$ yd., and $1\frac{3}{4}$ yd.?

22. A clerk sold $27\frac{3}{4}$ yd. of cotton goods to one lady and $41\frac{7}{8}$ yd. to another. How much has he left from a piece containing 100 yd.?

23. Find the cost of oil cloth at $1.20 a sq. yd. to cover a kitchen floor 12 ft. by 15 ft.

24. A wall is 50 ft. long and 36 ft. wide. Find the cost of plastering this wall at 55¢ a sq. yd.

25. A steamer traveled 6699 miles in 319 hours. Find the average rate of speed per hour.

26. Find how much is gained on 8380 yd. of cloth bought at 11¢ and sold at 19¢ a yd.

27. By selling a ship for $83,475.50 a company lost $9460.25. How much did the ship cost?

28. A dealer invested $9463.32 in farm land. The house and barn cost him $\frac{4}{9}$ as much as the land. How much money did he invest in all?

29. An automobile started on a trip of 975 miles. After going $\frac{4}{5}$ of the distance, the driver found he had no more gasoline. How many miles had he still to go?

30. On the trip the machine used 75 gallons of gasoline. How many miles did the driver get out of each gallon?

31. At $4.35 each, what will 152 tables cost?

32. Find the number of square yards in a school area 26 yd. long and $10\frac{1}{2}$ yd. wide.

33. A farm was sold for $8450. The purchaser paid $\frac{3}{5}$ of the price in cash. How much cash did he pay?

34. At $9.62 each, what will 8 dozen girls' suits cost?

35. A train ran $47\frac{2}{3}$ mi. to its first stop, $86\frac{1}{2}$ mi. to the second stop, and $61\frac{5}{8}$ mi. to the last stop. How far did it travel?

TABLES FOR REFERENCE

Length

12 inches (in.)	=1 foot (ft.)
3 feet	=1 yard (yd.)
5½ yards	=1 rod (rd.)
16½ feet	=1 rod
320 rods	=1 mile (mi.)
1760 yards	=1 mile
5280 feet	=1 mile

Liquid Measure

2 pints (pt.)	=1 quart (qt.)
4 quarts	=1 gallon (gal.)

Dry Measure

2 pints (pt.)	=1 quart (qt.)
8 quarts	=1 peck (pk.)
4 pecks	=1 bushel (bu.)

Time

60 second (sec)	=1 minute (min.)
60 minutes	=1 hour (hr.)
24 hours	=1 day (da.)
7 days	=1 week (wk.)
4 weeks (about)	=1 month (mo.)
365 days	=1 common year
366 days	=1 leap year
12 months	=1 year
100 years	=1 century

Thirty days has September,
April, June, and November,
All the rest have thirty-one,
Excepting February alone,
To which we twenty-eight assign,
Till leap year gives it twenty-nine.

Weight

16 ounces (oz.)	=1 pound (lb.)
100 pounds	=1 hundredweight (cwt.)
2000 pounds	=1 short ton (T.)
2240 pounds	=1 long ton (L. T.)

Counting

12 things	=1 dozen (doz.)
12 dozen	=1 gross
20 things	=1 score

Paper Measure

24 sheets	=1 quire
20 quires	=1 ream

Square Measure

144 square inches (sq. in.)	=1 square foot (sq. ft.)
9 square feet	=1 square yard (sq. yd.)

Cubic Measure

1728 cubic inches (cu. in.)	=1 cubic foot (cu. ft.)
27 cubic feet	=1 cubic yard (cu. yd.)

MULTIPLICATION TABLES

2×1=2	3×1=3	4×1=4	5×1=5
2×2=4	3×2=6	4×2=8	5×2=10
2×3=6	3×3=9	4×3=12	5×3=15
2×4=8	3×4=12	4×4=16	5×4=20
2×5=10	3×5=15	4×5=20	5×5=25
2×6=12	3×6=18	4×6=24	5×6=30
2×7=14	3×7=21	4×7=28	5×7=35
2×8=16	3×8=24	4×8=32	5×8=40
2×9=18	3×9=27	4×9=36	5×9=45
2×10=20	3×10=30	4×10=40	5×10=50
2×11=22	3×11=33	4×11=44	5×11=55
2×12=24	3×12=36	4×12=48	5×12=60
6×1=6	7×1=7	8×1=8	9×1=9
6×2=12	7×2=14	8×2=16	9×2=18
6×3=18	7×3=21	8×3=24	9×3=27
6×4=24	7×4=28	8×4=32	9×4=36
6×5=30	7×5=35	8×5=40	9×5=45
6×6=36	7×6=42	8×6=48	9×6=54
6×7=42	7×7=49	8×7=56	9×7=63
6×8=48	7×8=56	8×8=64	9×8=72
6×9=54	7×9=63	8×9=72	9×9=81
6×10=60	7×10=70	8×10=80	9×10=90
6×11=66	7×11=77	8×11=88	9×11=99
6×12=72	7×12=84	8×12=96	9×12=108

10×1=10	11×1=11	12×1=12
10×2=20	11×2=22	12×2=24
10×3=30	11×3=33	12×3=36
10×4=40	11×4=44	12×4=48
10×5=50	11×5=55	12×5=60
10×6=60	11×6=66	12×6=72
10×7=70	11×7=77	12×7=84
10×8=80	11×8=88	12×8=96
10×9=90	11×9=99	12×9=108
10×10=100	11×10=110	12×10=120
10×11=110	11×11=121	12×11=132
10×12=120	11×12=132	12×12=144

Spotted something that isn't quite right?

Let us know and we'll fix it for future reprints. And as a thank you for your diligence, we'll send you a free Sothis book of your choice.

publisher@sothis.press

This page intentionally left blank

Made in United States
Orlando, FL
05 November 2026

90231725R00194